ALLERGIES AND INFECTIOUS DISEASES

# PEANUTS

# BIOACTIVITIES AND ALLERGIES

## ALLERGIES AND INFECTIOUS DISEASES

Additional books in this series can be found on Nova's website under the Series tab.

Additional E-books in this series can be found on Nova's website under the E-book tab.

## FOOD SCIENCE AND TECHNOLOGY

Additional books in this series can be found on Nova's website under the Series tab.

Additional E-books in this series can be found on Nova's website under the E-book tab.

ALLERGIES AND INFECTIOUS DISEASES

# PEANUTS

# BIOACTIVITIES AND ALLERGIES

IVÁN LÓPEZ EXPÓSITO
AND
ANA BELEN BLAZQUEZ
EDITORS

nova publishers
*New York*

For permission to use material from this book please contact us:
Telephone 631-231-7269; Fax 631-231-8175
Web Site: http://www.novapublishers.com



**Library of Congress Cataloging-in-Publication Data**

ISBN: 978-1-62417-387-5
Library of Congress Control Number: 2012952789

*Published by Nova Science Publishers, Inc. † New York*

# Contents

# PREFACE

Peanut is an important crop grown on over 21 million ha worldwide. World peanut production is approximately 34 million tons per year, being China the largest producer followed by India and the U.S. Although peanuts have had an adverse image due to their content in saturated fatty acids, it is important to mention they also are an excellent source of protein, vitamins and minerals for human nutrition. Furthermore their oil is widely consumed in Asian and tropical countries. In addition, peanut contains relevant bioactive constituents that can provide a wide range of health benefits. Relevant attributes of peanuts include relatively high levels of prebiotic low-digestive carbohydrates and significant levels of phytochemicals, such as resveratrol, phytosterols or phyto-oestrogens. Unfortunately, peanut is also one if not the most allergenic vegetables and the most common food-related cause of IgE-mediated allergic reactions, mainly in westernized countries. This kind of reactions might be life threatening in some subjects and require a deep and intensive research labor.

The book has been divided into two parts with four chapters each. In the first part, an overview regarding peanut composition, physical characteristics as well as the biological activity of peanut phytochemicals have been included. In Part 2, the chapters are focused on peanut allergy. The different chapters cover peanut and legumes allergens characteristics, together with their prevalence. They also described the therapeutic strategies to treat peanut allergy and the problem regarding the cross-reactivity of peanut and other allergens which is critical for the development of effective strategies to diagnose and treat peanut allergy and eventually to improve patient health.

In the book we have tried to include authors belonging to both research centers and hospitals where patients with food allergy problems are constantly observed. By this approach we would like to cover similarly basic and clinical aspects of the current research related with peanut. We hope the reading of this book will help you to better understand the main bioactive properties of peanut as well as the immunological mechanisms responsible for peanut allergy.

*Iván López Expósito*
*Ana Belén Blázquez*

# Part 1.
# Peanut Composition and Biological Activity

In: Peanuts: Bioactivities and Allergies
Editors: I.L. Expósito, A.B. Blázquez
ISBN: 978-1-62417-387-5


*Chapter 1*

# AN OVERVIEW OF PEANUTS AND THEIR BY-PRODUCTS: PRODUCTION AND CHEMICAL COMPOSITION

***Y. Aguilera, V. Benítez, R. M. Esteban, F. J. López-Andréu, E. Mollá and M. A. Martín-Cabrejas****
Instituto de Investigación en Ciencias de la Alimentación (CIAL)
CSIC-UAM/Departamento de Química Agrícola,
Facultad de Ciencias, Madrid, Spain

## ABSTRACT

Peanut (*Arachis hypogaea* L.) belongs to the legume family (*Fabaceae*) and produces small yellow flowers and subterranean pods (from 3 to 7 cm) in addition to the fruits, which contain 1 to 4 edible seeds. Cultivated peanut is an important source for oil and food crop and is currently grown on approximately 21 million ha worldwide. Although peanut was probably first cultivated in Peru, China is the world's largest peanut producer (42%) followed by India (18%) and USA (10%). There are four classes of peanuts in the market: Virginia, Runner, Spanish, and Valencia. A large part of peanut production is used for human

* E-mail address: maria.martin @uam.es, Instituto de Investigación en Ciencias de la Alimentación (CIAL) CSIC-UAM/Departamento de Química Agrícola, Facultad de Ciencias, Campus de la Universidad Autónoma de Madrid, 28049 Madrid, Spain.

consumption, but is also employed as a raw ingredient for processed foods such as peanut butter, confections, roasted peanuts and snack products. Furthermore it produces oil and extenders in meat product formulations, soups and desserts. In fact, as an oil seed crop, peanut ranks fourth in world production. Nowadays, the incorporation of high-oleic genes into new peanut breeding lines is being carried out, and these peanut varieties have a favorable high oleic acid content (80% of the total of fatty acids) and consequently extended shelf life, including a better resistance to several diseases. In addition, peanut contains important nutrients and bioactive constituents that can provide a wide range of health benefits. Relevant attributes of peanut include protein with very high arginine content, relatively high levels of insoluble and soluble dietary fiber, healthy fatty acids, besides to calcium, magnesium, and vitamins including important contents of folate, niacin and vitamin E. Thus, this food is made up of 49-60% oils and 27-30% protein, being an excellent source of energy (581 Kcal/ 100 g). Other important bioactive compounds present in peanuts are antioxidant polyphenols, resveratrol and phytosterols in relevant levels. The value of peanut components and their potential effects in the diet have been characterized and quantified; however, substantial amounts of by-products are generated in the process of peanut harvest and peanut oil extraction as unfriendly peanut wastes. At present, there is a great interest to recover the nutritional value of peanut by-products because they would provide new ingredients to the food industry and could represent a significant economic and social benefit.

## INTRODUCTION

Cultivated peanut, also called groundnuts, was originated in South America. Evidence for the earliest known use of peanuts comes from an archaeological site in Peru and dates 3,500 years ago [1]. To date, eighty different species belonging to the genus *Arachis* have been described with the exception of some primitive land races in South America. However, only *A. hypogaea* L. is grown for human consumption. With the exploration of the New World at the beginning of 16$^{th}$ century, peanut quickly spreads out from South America to Africa and Asia along with other new world crops such as bean, squash, potato, tobacco and corn [1]. Its distribution occurred in two ways whereby two seeded types of Brazilian origin were taken to Africa, and three seeded types of Peruvian origin found their way to the west coast of South America, China and various islands in the western Pacific. It was later in the 18$^{th}$ century, "Spanish" types were taken to Europe where they were grown for oil and human consumption. Introductions to North America were presumed to have been from northern Brazil or the West Indies, where they were loaded onto ships carrying slaves from Africa to the New World as food supplies [1]. Peanut is cultivated around the world in tropical, sub-tropical and temperate climates.

## BOTANICAL DESCRIPTION

Peanut is not a true nut, belongs to the *Fabaceae* family which is comprised of the leguminous plants. Groundnut is morphologically variable and there are many recognizably distinct varieties. The species and varieties are classified according to the location of the flowers on the plant, patterns of reproductive nodes on the branches, number of trichomes, as well as pod morphology [2]. Peanut is a member of the genus *Arachis*, which poses a unique characteristic that distinguishes it from all other plants that it is flowering above ground and producing fruits below ground, hence the name peanut [3]. Variants of *A. hypogaea* have frequently been described as distinct species, subspecies and botanical varieties. *A. hypogaea* is divided into two subspecies, *hypogaea* (alternately branched) and *fastigiata* (sequentially-branched) [4] (Figure 1). The *hypogaea* subspecie includes the *hirsuta* (Runner type) and *hypogaea* (Virginia type) varieties, which are divided into variety groups according to pod characters, seed numbers and general appearance, and subsequently into variety clusters based on habit. The *fastigiata* subspecie (all upright forms) includes the *fastigiata* (the very distinct 'Valencia' forms) and *vulgaris* (the 'Spanish-Natal-Manyema' complex) varieties, separated on habit and characters of the inflorescence. These varieties are further divided, according to pod and seed characters, into variety groups and clusters. Finally, as in the *hypogaea* subspecie, the individual varieties may be distinguished according to testa colour [5].

Peanut plant is an annual/perennial herb, with an undetermined mode of growth [6]. A peanut seed consists of two cotyledons, a stem axis and leaf primordial, hypocotyls and primary root. Cultivated peanut is generally self-pollinating although out-crossing does occur in less extent. The wild peanut relatives are also assumed to be self-pollinating. The geocarpic habit of peanut is a unique characteristic and could be responsible for dispersal and thus, population structure [7]. Day length is crucial in the peanut's development as the plant produces more flowers under long day conditions, but reproductive efficiency is higher during short days. The main difference between annual and perennial peanut plants is the morphology of the stigma. In perennial species, the stigma would be covered with hair in order to hinder pollen attachment. Fertilization would occur after two days pollination, and the ovary would elongate geotropically to form a peg. In domesticated *A. hypogaea* the peg might elongate for up to two days after soil penetration while in the wild species, it may continue to elongate for up to meters [8]. Regarding to peanut seed germination, it is of epigeal mode and the cotyledons tend to change colour to green after emergence. The primary root system of peanut is a tap root system and numerous lateral roots are visible on the third day after germination [9]. Peanut roots do not have the normal root hair, but tuffs of hair can be seen on the lateral root. The former is only restricted to a root zone of 35 cm below the soil surface. Root hairs generally serve as invasive sites in most legumes, but this is not the case in peanut [10].

Peanut could be affected by many diseases including early and late leaf spot, southern stem rot, *Cylindrocladium* black rot, web blotch, tomato spotted wilt virus, nematodes, and *Sclerotinia* blight. The latter caused by *Sclerotinia minor Jagger* is a major disease of peanut and other crops, such as common bean, and lettuce. *Arachis* has little natural resistance to this pathogen [11]. The necrotrophic pathogen produces cottony white mycelia and oxalic acid which predispose the plant to invasion by the fungus.

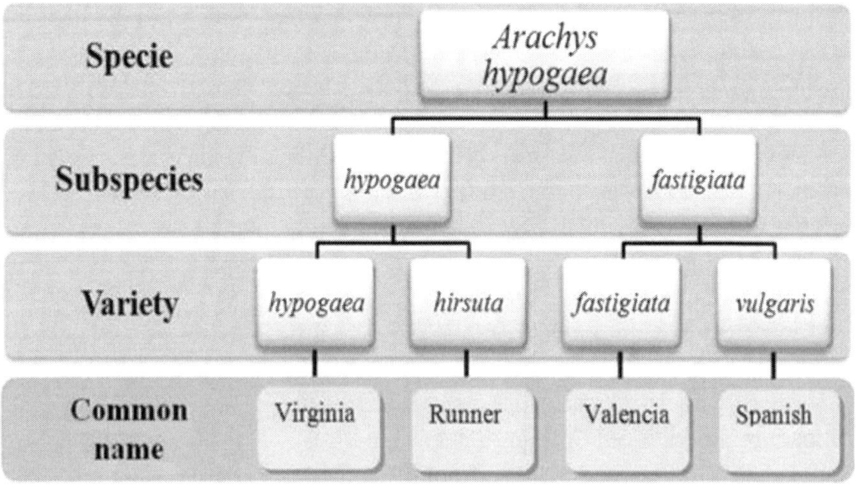


Figure 1. The classification of *Arachys hypogaea* L. varieties.

## CULTIVARS

The most popular peanut cultivars are Spanish, Runner, Virginia, and Valencia. Although they show differences in flavor, oil content, size, shape, and disease resistance, these cultivars can be interchangeable for particular uses. In general, Virginia-type and also some Valencia-type peanuts are marketed by the shell. Spanish-type peanuts are mostly preferred for peanut candies, salted nuts, and peanut butter. Runner-type peanuts are also used for peanut butter.

### Runner

Runners have become the dominant peanut type grown in the U.S. (80% of total production) due to the introduction in the early 1970's of a new variety (Florunner), which was responsible for a spectacular increase in peanut yields. Runner-type peanuts have rapidly gained wide acceptance due to their attractive kernel size range. A high proportion of these peanuts are used for peanut butter because they show good flavor, better roasting characteristics and higher yields when compared to Spanish-type peanuts.

Cultivars of Runners include "Southeastern Runner 56-15", "Dixie Runner", "Early Runner", "Virginia Bunch 67", "Bradford Runner", "Egyptian Giant", "Rhodesian Spanish Bunch", "North Carolina Runner 56-15", "Florunner", "Virugard", "Georgia Green", "Tamrun 96", "Flavor Runner 458", "Tamrun OL01", "Tamrun OL02" "AT-120", "Andru-93", "Southern Runner", "AT1-1", "Georgia Brown", "GK-7" and "AT-108" [4].

### Virginia

Virginia-type peanuts grow either bunch or running in growth habit. The bunch type is upright to spreading, reaches a height of 45 to 55 cm, and a spread of 70 to 80 cm, with 80 to 90 cm rows that seldom cover the ground. The pods are borne within 5 to 10 cm of the base of the plant. Virginias exhibit the largest kernels and account for most of the peanuts roasted and eaten as "inshells." When shelled, the larger kernels are sold as salted peanuts. Cultivars of Virginia-type peanuts include “NC 7”, “NC 9”, “NC 10C”, “NC-V 11”, “VA 93B”, “NC 12C”, “VA-C 92R”, “Gregory”, “VA 98R”, “Perry”, “Wilson”, “Hull”, “AT VC-2” and “Shulamit” [4].

### Spanish

Spanish-type peanuts have smaller kernels covered with a reddish-brown skin. They contain higher oil level than the other types of peanuts and are mostly used in peanut candies and peanut butter. Cultivars of the Spanish group include "Dixie Spanish", "Improved Spanish 2B", "GFA Spanish", "Argentine", "Spantex", "Spanette", "Shaffers Spanish", "Natal Common", "White Kernel Varieties", "Starr", "Comet", "Florispan", "Spanhoma", "Spancross", "OLin", "Tamspan 90", "AT 9899-14", "Spanco" "Wilco I", "GG 2", "GG 4", "TMV 2", and "Tamnut 06".

### Valencia

Valencia-type peanuts have usually three or more small kernels to a pod and they are comparatively tall, having a height of 125 cm and a spread of 75 cm. Peanut pods are borne on pegs arising from the main stem and the side branches. Most of the pods are clustered around the base of the plant. Valencia types are three to five seeded and smooth, with no constriction of the shell between the seeds. Seeds are oval and tightly crowded into the pods. These peanuts are very sweet and mainly used for sale roasted and salted, in-shell peanuts and peanut butter. Cultivars of Valencias include “Valencia A” and “Valencia C”.

## Peanut Production

Peanut is an important crop grown on over 21 million ha worldwide (Table 1). World peanut production is approximately 34 million tons per year, being China the largest producer followed by India and the U.S. [12]. Worldwide peanut exports are approximately 1.25 million tons, being the U.S. one of the world’s leading peanut exporters, with average annual exports between 200,000 and 250,000 tons. Canada, Mexico and Europe account for over 80% of U.S. exports. The largest US export markets within Europe are the Netherlands, the U.K., Germany, and Spain. Argentina and China are also significant exporters; countries such as India, Vietnam and several African countries periodically enter the world market depending upon their crop quality and world market demand.

Although competition within a dynamic snack market continues to put pressure on peanuts to compete with a growing range of products (potato chips, extruded snacks, tree nuts, and baked snacks), demand for peanuts in North America and Europe has been steady. Quality specifications, food safety concerns and import requirements continue to require the implementation of improved monitoring and quality control standards at origin.

**Table 1. World-wide peanut area, production, oil production and protein meal production in 2000/02-2010/12**

| Year | 2000 | 2002 | 2004 | 2006 | 2008 | 2010 |
|---|---|---|---|---|---|---|
| Peanut area (Million hectares) | 23 | 23 | 24 | 22 | 23 | 21 |
| Peanut production (Million metric tons) | 31 | 31 | 33 | 31 | 34 | 34 |
| Production of peanut oil (Million metric tons) | 4 | 5 | 5 | 5 | 5 | 5 |
| Peanut protein meal production (Million metric tons) | 5 | 5 | 6 | 5 | 6 | 6 |

Adapted from Zhao et al. (2011) [49].

## Peanut Composition

Peanut as a leguminous seed can make an important contribution to the human diet in many ways. It is a good source of protein, lipid, and fatty acids for human nutrition [13-19] (Table 2). In addition, peanut contains relevant bioactive constituents that can provide a wide range of health benefits. Relevant attributes of peanuts include relatively high levels of prebiotic low-digestive carbohydrates such as insoluble and soluble dietary fiber, and significant levels of phytochemicals, such as resveratrol, phytosterols or phyto-oestrogens.

### Protein

Peanut is considered an excellent source of protein, ranging from 27 to 30% [20]. Peanut proteins constitute an interesting alternative due to their high nutritional value, functional properties and low cost. The latter is critically needed in many developing countries, because animal protein is more expensive and is getting beyond the reach of many people. The amino acid profile of the peanut meals shows that peanut might be an ingredient for protein fortification [21]. The major amino acid found in peanut is arginine, which is a precursor to nitric oxide that helps expand the blood vessels. Actually, it has been reported that peanut consumption improves indicators of cardiovascular disease risk in healthy adults [22-23]. Arginine besides to aspartic acid and glutamic acid may account for about 45% of the total amino acid in peanut seeds. The limiting amino acids, methionine, lysine, and threonine are usually ranging from 0.8 to 1.2% [20, 24]. Nevertheless, the content of amino acids in peanut seeds varies according to genotype, variety, growing season, environmental factors, and maturity [25]. It has been demonstrated that an $CO_2$-enriched environment at 700 or 1400 ppm did not appear to significantly affect the amino acid profile of peanut seeds except for methionine. However, that difference is only 2-3 mg/g of protein at 700 ppm and may not be meaningful [24].

**Table 2. Nutritional composition of peanut products***

| | Peanut seed | Roasted peanut | Butter peanut |
|---|---|---|---|
| Protein (g/100g) | 25-30 | 24-30 | n.d. |
| Arginine (g/100g of protein) | 12-14 | 12-14 | n.d. |
| Aspartic acid (g/100g of protein) | 10-13 | 11-14 | n.d. |
| Glutamic acid (g/100g of protein) | 20-22 | 21-23 | n.d. |
| Carbohydrate (g/100g) | 19-30 | 17-25 | n.d. |
| Dietary fiber (g/100g) | 4.1-4.8 | 4.3-5.0 | 18-24 |
| Fat (g/100g) | 31-51 | 46-50 | 16-21 |
| Saturated fatty acids (g/100g of total fatty acids) | 18-25 | 7-17 | 27-35 |
| Polyunsaturated fatty acids (g/100g of total fatty acids) | 25-47 | 25-31 | 48-55 |
| Oleic acid (g/100g of total fatty acids) | 30-46 | 27-36 | 26-32 |
| Linoleic acid (g/100g of total fatty acids) | 29-41 | 21-29 | 1.2-1.7 |
| Oleic/Linoleic acid | 0.6-1.3 | 0.9-1.3 | 1.8-2.2 |
| Ash (g/100g) | 2.5-3.0 | 1.9-2.3 | n.d. |
| Potassium (mg/100g) | 663-867 | 687-734 | n.d. |
| Mangesium (mg/100g) | 162-223 | 178-183 | n.d. |
| Phosphorus (mg/100g) | 225-290 | 288-348 | n.d. |
| α-Tocoferol (mg/100g) | 6.8-10.2 | 3.2-7.2 | 50-55 |
| Folate (mg/100g) | 95-106 | 63-69 | n.d. |
| Total Phytosterol (mg/100g) | 113-129 | 85-91 | n.d. |
| β-sitosterol (mg/100g) | 66-77 | 44-52 | n.d. |
| *trans*-resveratrol (μg/g) | 0.09-0.60 | n.d. | n.d. |

* Khalil and Chughtai [61]; Hinds [62]; Wu et al. [24]; Grosso et al. [63]; Ozcan and Seven [64]; Isleib et al.[17]; Shin et al. [18]; Rodriges et al. [65].

n.d.= not determinate

Peanut proteins show good emulsifying activity, emulsifying stability, foaming capacity, excellent water retention and high solubility, and they also provide a new high protein food ingredient for product formulation and protein fortification in food industry [26]. Thus, the peanut proteins can be regarded as one of the most attractive and promising vegetable proteins. These properties are affected by the intrinsic factors of protein such as molecular structure and size, and many extrinsic factors including the method of protein separation/production, pH, ionic strength and the presence of other components in the food system.

## Fatty Acid Composition

Average oil content of peanut seed may reach 50% of total weight [27, 20]. The fatty acid composition contains palmitic (16:0), stearic (18:0), oleic (18:1), linoleic (18:2), arachidic (20:0), eicosenoic (20:1), behenic (22:0), and lignoceric (24:0) acids, being oleic and linoleic acids the principal components (80% of the total fatty acids). It has been demonstrated that maturity and environmental factors, including temperature and moisture, could significantly affect the ratio of oleic acid to linoleic acid in peanut oil. The effect of increasing temperature on peanut oil composition promoted an increase of monounsaturate acid levels beside to a decrease of polyunsaturate acid contents [28]. An increase of linoleic acid content and a decrease of oleic acid content were reported for dry land grown peanuts when supplemental irrigation was used [29-30]. Other environmental factors such as photoperiod might also be involved and may be worth further investigation [24].

The oleic to linoleic acid ratio is a quality index employed for the determination of genetic peanut characteristics classified as normal, mid-, and high-oleic types [18]. In previous studies involving multiple peanut cultivars, a strong negative correlation between oleic acid and linoleic acid in peanut lipids was also observed [31]. This negative relationship is originated from the biochemical pathways of peanut development; palmitoyl CoA is elongated to stearoyl CoA followed by desaturation forming oleic acid in the peanut germplasm.

Then, the action of oleoyl-phosphatidylcholine desaturase (a D12-fatty acid desaturase) synthesizes linoleic acid from oleic acid [32-33].

In addition, palmitic acid is positively correlated to linoleic acid ($r = 0.967$, $P < 0.001$) [31], while negatively correlated to oleic ($r = -0.971$, $P < 0.001$) and gondoic acids ($r = -0.860$, $P < 0.001$) [34]. Nowadays, the incorporation of high-oleic genes into new peanut breeding lines is carried out, and these peanut varieties have a favorable high oleic acid (80% of the total of fatty acids) and consequently extended shelf life, including a better resistance to several diseases [35].

## Carbohydrate Fraction

Sugars made by photosynthesis in the leaves move into the seeds where they are converted into different energy-containing constituents. During early development, the peanut seed converts the incoming sugars primarily into starch, and during later maturation, the sugars are converted primarily into oil [17].

The content of carbohydrates is 30-32% and includes dietary fiber, starch, pentosans, and sugars [24]. The main carbohydrates detected in peanut are sucrose, ranging from 3.0 to 10.0 mg/g, with the occasional presence of fructose and glucose. On the other hand, some α-galactosides have been also identified as raffinose and stachyose [36]. Dietary fiber average is 6 g/100g, being 1.9 g soluble fiber. It may contribute to the fact that peanut has a low glycemic index and glycemic load [37]. Soluble fiber reduces total and LDL-cholesterol and plays an important role in guarding against bowel disease. In addition, the relevant dietary fiber content of peanuts may in part contribute to the satiating properties of peanuts, compared to other snack foods [38]. Vercellotti et al. [39] studied peanut mono-, oligo-, and polysaccharide fractions as the origin of intermediates for flavor molecules. These fractions of carbohydrates contribute to sensory peanut quality during roasting process. In this sense, possible relationships between carbohydrate components in peanuts and selected sensory attributes (sweet, bitter, and roasted) were found [36]. Positive correlations between total sugars and sweetness ($P < 0.05$), and roasted peanut attribute were shown, and a negative correlation between total sugars and bitterness was also exhibited. Glucose and fructose contents were associated with the roasted peanut attribute. Raffinose content had a significant negative correlation with astringency ($r = -0.955$, $P < 0.05$) and, similar to glucose and fructose contents, was positively correlated with roasted peanut attribute. In contrast, stachyose content was negatively correlated with bitterness. The individual components of the peanut carbohydrate fractions changed during maturation, curing, and they could also change due to climatic variances in geographical regions, storage time and conditions, or the dry roasting process [40].

## Minerals and Vitamins

The mineral profile of peanut seeds shows that potassium, phosphorus, magnesium, and calcium are the predominant mineral elements. Peanut can potentially contribute a significant proportion of the daily requirement for copper and magnesium, if regularly consumed in modest amounts. 100 g of roasted peanuts can furnish approximately 100% of Cu, 62% of Mn, 60% of Mg, 39% of P and 33% of Zn of the recommended dietary allowance (RDA).

Moreover, peanut is one of the best dietary sources of vitamins, including vitamin E, folate, niacin, thiamin, riboflavin and pantothenic acid. Values of α-tocoferol of 10.5 mg/100 g were determined in Runner-type peanuts. An ounce of peanuts provides 20% of our daily needs and thus, it is considered an excellent source of vitamin E. However, published data on the vitamin E content of raw peanuts are variable and difficult to interpret. Most studies reporting vitamin E levels in raw peanuts have been based on experimental growth conditions that do not reflect commercial production using present-day agronomic practices. The combination of analytical data with chemometric techniques might be used by peanut breeders to give more information on peanut cultivars than experimental data alone [41]. Therefore, the consuming of peanuts and peanut-containing foods on a daily basis is a healthy choice in trying to reach the RDA for vitamin E. Peanut is also a good source of folate providing 110μg/100g in plain peanuts, 66μg/100g in roasted peanuts, and 53μg/100g in peanut butter. It has also been suggested that regularly consuming nuts may favorably influence homocysteine levels, by reducing them.

## Flavors

Volatile components of raw and processed peanuts have been identified. Heterocycles (pyrazines and pyrroles), sulfur compounds, phenols, ketones, esters, alcohols, and hydrocarbons have been detected [42]. Oxazoles, thiazoles, and oxazolines have been also shown [43-44]. Although no correlation between oil content and flavor has been reported in literature, it might be expected that altering oil content could have an impact on peanut flavor. The quality of peanuts determined by chemical composition and flavor is very important to their acceptability, nutritional and health benefits to consumers [45].

### Phytochemicals

Phytochemicals are plant chemicals produced by plants to protect them from diseases and attacks. Peanuts and peanut products contain many different bioactive phytochemicals including several flavonoids, such as luteolin, quercetin, kaempferol, rutin; isoflavones (phyto-oestrogens); phenolic compounds, such as ellagic acid and resveratrol; lignans; and phytosterols, in particular, β-sitosterol. They can provide a wide range of health benefits such as reducing coronary heart diseases and cancer [46-48]. Anyhow, the will be discussed deeply, in another chapter of this book.

## PEANUT BY-PRODUCTS

The food industry generates a large amount of by-products in the process of peanut harvest and peanut oil extraction, making it necessary to search possible ways for their utilization. Only few of these by-products are used as animal feeding and organic fertilizers.

During the last 20 years, new scientific research is focused on the nutrient and bioactive constituents, their potential effects in the diet and their effects on human health. At present, the isolation and purification of bioactive compounds from peanut processing by-products, like peanut meals, skins, hulls and vines are being studied [49]. These bioactive compounds can be processed by separation and purification techniques or mechanical engineering and then being used as food ingredients rich in specific compounds.

Examples of bioactive compounds present in such wastes are *t*-resveratrol, procyanidins, dietary fiber and high value protein. Although there are previous studies on the functional components and properties in peanut meals, skins, hulls and vines, commercial products from by-products are absent.

### Peanut Meal

From the oil mills, it comes a large peanut by-product known as peanut cake or meal, which is well separated from peanut kernel after oil extraction. The average production of peanut meal all over the world is 6 million tons in 2010 (Table 1). The peanut meal is an inexpensive and underutilized by-product of the peanut industry which contains 50-55% high quality protein

[49]. Despite having an excellent amino acid profile, it has only found marginal use in food industry due to its inferior functional properties. The development of by-products made from peanut meal could provide a high protein food ingredient for formulation and protein fortification in food products [26]. Moreover, there are other compounds in significant quantities that may impact functionality of peanut protein, such as polysaccharides, which reduce *in vivo* and *in vitro* protein digestibility and nutrient absorption due to the interactions between these two macromolecules [50].

Nowadays, the isolation and enrichment of *t*-resveratrol and other polyphenols from peanut press waste is carried out using molecular imprinting solid phase extraction technology, which provides an efficient and sustainable path to use peanut wastes [51].

## Peanut Skin

Peanut skin is a low-value by-product of peanut processing operations. An estimated 35-45 g of peanut skin is generated per kg of shelled peanut kernel. Over 0.74 million tons of peanut skins are annually produced worldwide as by-products of peanut processing industry [52]. Natural phenolic compounds can be extracted from peanut skins, showing high antioxidant activity [53]. Three classes of phenolic acids (caffeic acid, chlorogenic acid, ferulic acid and coumaric acid), flavonoids (catechins and procyanidins), and stilbene (resveratrol) have been reported [54]. In addition, several proanthocyanidins and ethyl protocatechuate have been identified in peanut skins, showing significant antioxidant activities [41]. Thus, peanut skin can provide an inexpensive source of polyphenols to use them as functional ingredients in foods or dietary supplements [55]. In this sense, natural phenolic compounds extracted from peanut skin and hull were used as source of natural antioxidants in meat applications, observing an increase in the quality and storage stability of beef products [56].

## Peanut Hull

The production of peanut hull has been estimated to be 230-300 g of peanut hull per kg of peanut. Most peanuts are separated from the shells or hulls prior to processing into value added products. The hulls, which are biodegradable and absorbent, can be utilized as animal food fillers, absorbents, or carriers for pesticides or fertilizers, although they are often landfilled. Interestingly, peanut hulls are a good source of dietary fiber and other bioactive components; therefore they are gaining status as a useful commodity. The peanut hulls consist of some constituent polymers: cellulose, hemicelluloses, pectin, lignin and protein. Milling and fractionation of peanut hull can isolate important dietary fiber components for incorporation into commercial food products to enrich their fiber content and/or serve as functional ingredients. Expanding peanut hull utilization in food products can serve to enhance human health, and increases the market value of the crops [57]. Recent studies have focused on utilizing fibers from peanut hull for use in nonwoven fabrics. The results demonstrated that this peanut fiber obtained from the hulls, was weak yet stiff, in comparison with other textile fibers and may be useful in certain geotextile applications [58].

### Peanut Vines

Roots, stems, leaves and flowers are part of the peanut vines. It has been estimated 60-65% of peanut vine production respect to total peanut production. Although few information is available related to the bioactive components of peanut vines, it is known that it contain high levels of phenolic compounds and dietary fiber [59]. Flavonoid compounds, such as rutin, quercetin and luteolin are also present in peanut vines [60], providing a wide range of health benefits.

## Conclusion

Peanut (*Arachis hypogaea* L.) is one of the most important oil crops in the world with a world production of approximately 34 million tons per year. The main cultivars are Spanish, Virginia, Valencia and Runner types. Peanut is a good source of nutritional and bioactive compounds (dietary fiber, protein, arginine, unsaturated acids, α-tocoferol, magnesium and potassium, resveratrol and sterols) which may have important metabolic and/or physiological effects. More recent evidence, provides potential information of its impact on health, and its possible uses as ingredients on functional foods. Although peanut presents many uses for human consumption, such as raw legume, peanut oils, peanut butter, or roasted peanut, substantial amounts of by-products are generated. Nowadays, many studies have been carried out to investigate the possible functional components and properties in peanut by-products such as peanut meals, skins, hulls and vines, providing an efficient and sustainable path to use peanut wastes. Future research is focused on studying the potential of these by-products as food or feed additives using modern technologies, and in revaluating their effects on human health.

## REFERENCES

[1] R.O. Hammons, In J Smartt, ed, *The Groundnut Crop: A scientific basis for improvement.* Chapman & Hall, London, 24 (1994).
[2] Krapovickas and W.C. Gregory, *Bonplandia* 8, 1 (1994).
[3] C.C. Holbrook and H.T. Stalker, In J Janick, ed, *Plant Breeding Reviews,* Vol 22. John Wiley and Sons, Inc., 297 (2003).
[4] A.K. Singh and C.E. Simpson, In J Smartt, ed, *The Groundnut Crop: A scientific basis for improvement.* Chapman & Hall, London, 96 (1994)
[5] R.A. Gibbons, A.H. Bunting and J. Smart, *Euphytica* 21, 78 (1972).
[6] H.T. Stalker, *Field Crop. Res.* 53, 205 (1997).
[7] G. Kochert, T. Halward, W.D. Branch, and C.E. Simpson, *Theor. Appl. Genet.* 81, 565 (1991).
[8] J. Lu, A. Mayer, and B. Pickersgil, *Ann. Bot.* 66, 73 (1990).
[9] W.C. Gregory, M.P. Gregory, A. Krapovickas, B.W. Smith and J.A. Yarbrough, *Am. Peanut Res. Educ. Soc. Stillwater*, 47 (1973).
[10] J.P. Moss and V.R. Rao, In: Patee, H.E., and Stalker, H.T. (ed), Advances in peanut science. *Am. Peanut Res. Educ. Soc, Stillwater,* 1 (1995).
[11] D.M. Porter, H.A. Melouk and B. Sclerotinia, In Kokalis-Burelle N, Porter DM, Rodríguez-Kábana R, Smith DH, Subrahmanyam P eds, *Compendium of Peanut Diseases* 2nd edition. APS Press, St. Paul, 34 (1997).
[12] FAS-USDA Statistics (2011) Retrieved December 2011 from *http://www.fas.usda.gov/*
[13] E.M. Gaydou, J.P. Bianchini and J. Ratovogery, *J. Agric. Food Chem.* 31, 833 (1983).
[14] Y. Aguilera, M. Dueñas, I. Estrella, T. Hernández, V. Benítez, R.M. Esteban et al., *J Agric. Food Chem.* 58(18) 10101 (2010).
[15] Y. Aguilera, V. Benítez, E. Mollá, R.M. Esteban and M.A. Martín-Cabrejas, *Plant Foods Hum. Nutr.* 63, 391 (2011a).

[16] E.H. Ahmed and C.T. Young, In *Peanut Science and Technology*; Pattee, H. E., Young, C. T., Eds.; Am. Peanut Res. Educ. Soc. Inc.: Yoakum, TX; 655 (1982)

[17] T.G. Isleib, H.E. Pattee and F.G. Giesbrecht, *J. Agric. Food Chem.* 52, 3165 (2004).

[18] E.C. Shin, B.D. Craft, R.B. Pegg, R.D. Phillips and R.R. Eitenmiller, *Food Chem.* 219 1262 (2010a).

[19] K.L. Bett, J.R. Vercellotti, N.V., Lovegren, T.H. Sanders, R.T. Hinsch and G.K. Rasmussen, *Food Chem.* 51, 21 (1994).

[20] S.K. Pancholy, R. Sepulveda and S.M.M. Basha, *Proc. Am. Peanut Res. Educ. Soc.* 12, 13 (1980).

[21] J. Yu, M. Ahmedna and I. Goktepe, *Food Chem.* 103, 121 (2007)

[22] C.M. Alper and R.D. Mattes, *J. Am. Coll. Nutr. Apr.* 22(2), 133 (2003).

[23] J.F. Swain, P.B. McCarron, E.F. Hamilton, F.M. Sacks and L.J. Appel, *J. Am. Diet. Assoc.* 108(2), 257 (2008).

[24] W.H. Wu, J.Y. Lu, A.R. Jones, D.G. Mortley, P.A. Loretan, C.K. Bonsi et al. , *J. Agric. Food Chem.* 45, 3863 (1997).

[25] A.R. Hovis, C.T. Young and P.Y.P. Tai, *Peanut Sci.* 9, 44 (1982).

[26] H. Wu, Q. Wang, T. Ma and J. Ren, *Food Res. Int.* 42, 343 (2009).

[27] N.L. Derise, H.A. Lau, S.J. Ritchey and E.W. Murphy, *J. Food Sci.* 39, 264 (1974).

[28] D.F. Brown, C.M. Cater, K.F. Mattil and J.G. Darroch, *J. Food Sci.* 40, 1055 (1975).

[29] C.T. Young, M.E., Mason, R.S. Matlock, and G.R. Waller, *J. Am. Oil Chem. Soc.* 49, 314 (1972).

[30] C.T. Young, R.E. Worthington, R.O. Hammons, R.S. Matlock, G.R. Waller and R.D. Morrison, *J. Am. Oil Chem. Soc.* 51, 312 (1974).

[31] P.C. Andersen and D.W. Gorbet, *J. Agric. Food Chem.* 50, 1298 (2002).

[32] S. Jung, G. Powell, K. Moore and A. Abbott, *Molec. Gen. Genet.* 263, 806 (2000).

[33] M. Patel, S. Jung, K. Moore, G. Powell, C. Ainsworth and A. Abbott, *Theor. Appl. Genet.* 108, 1492 (2004).

[34] E.G. Hammond, D. Duvick, T. Wang, H. Dodo and R.N. Pittman, *J. Am. Oil Chem. Soc.* 74, 1235 (1997).

[35] R.S. Jonnala, N.T. Dunford and K.E. Dashiell, *J. Am. Oil Chem. Soc.* 82, 125 (2005).

[36] H.E. Pattee, T.G. Isleib, F.G. Giesbrecht and R. McFeeters, *J. Agric. Food Chem.* 48, 757 (2000).

[37] K. Foster-Powell, S.H.A., Holt and J.C. Brand-Miller, *Am. J. Clin. Nutr.* 76, 5 (2002).

[38] S.V. Kirkmeyer and R.D Mattes, *Int. J. Obes.* 24, 1167 (2000).

[39] J.R. Vercellotti, K.L. Bett and K.S. Choi, In 7th *International flavor symposium,* Samos, Greece; Charalambos, G., Ed.; Elservier Publishing: Amsterdam, The Netherlands, 1992; 232 (1972).

[40] J.R. Vercellotti, T.H. Sanders, S.Y. Chung, K.L. Bett and B.T. Vinyard, In *Food Flavors: generation, analysis and process influence*; Charalambos, G., Ed.; Elservier Publishing: Amsterdam, The Nerherlands, 1547 (1995).

[41] E.C. Shin, Y.Z. Huang, R.B. Pegg, R.D. Phillips and R.R. Eitenmiller, *J. Agric. Food Chem.* 57(21), 10289 (2009).

[42] J.P. Walradt, A.Q. Pittet, T.H. Kinlin, R. Muralidhara and A. Sanderson, *J. Agric. Food Chem.* 19, 972 (1971).

[43] M.H. Lee, C.T. Ho, and S.S. Chang, *J. Agric. Food Chem.* 29, 684 (1981).

[44] L.V. Burroni, N.R. Grosso and C.A. Guzmán, *J. Agric. Food Chem.* 45, 3190 (1997).

[45] K.L. Bett, J.R. Vercellotti, N.V. Lovegren, T.H. Sanders, R.T. Hinsch and G.K. Rasmussen, *Food Chem.* 51, 21 (1994).

[46] P.M Kris-Etherton, G. Zhao, A.E. Binkoski, S.M. Coval and T.D. Etherton, *Nutr. Rev.* 59(4), 103 (2001).

[47] L. Luo and Y.M. Huang, *Zhong Nan Da Xue Xue Bao Yi Xue Ban* 31(4), 566 (2006).

[48] T.A. Woyengo V.R. Ramprasath and P.J. Jones, *Eur. J. Clin. Nutr.* 63(7), 813 (2009).

[49] X. Zhao, J. Chen and F. Du, *J. Food Sci. Technol.* DOI 10.1007/s13197-011-0449-2. (2011).

[50] J. Mouécoucou, C. Villaume, C. Sanchez and L. Méjean, *Food Res. Int.* 37, 777 (2004).

[51] L.J. Schwarz, B. Danylec, Y. Yang, S.J. Harris, R.I. Boysen and M.T.W. Hearn, *J. Agric. Food Chem.* 59, 3539 (2011).

[52] V.S. Sobolev and R.J. Cole, *J. Sci. Food Agric.* 84, 105 (2003).

[53] Seo and C.V. Morr, *J. Food Sci.* 50(1), 262 (1985).

[54] J.M. Yu, M. Ahmedna and I. Goktepe, *Food Chem.* 90, 199 (2005).

[55] J.M. Yu, M. Ahmedna, I. Goktepe and J. Dai, *J Food Comp. Anal.* 19, 364 (2006).

[56] S.F. O'Keefe and H. Wang, *Meat Sci.* 73, 278 (2006).

[57] F.D. Oliveira, J.H. Paula, O.M. Freitas and S.A. Figueiredo, *Desalination* 248, 931 (2009).
[58] N. Bieakl and B.R. George, *Int. Nonw. J. Winter*, 45, 60 (2003).
[59] J.W. Dorner, R.J. Cole, T.H. Sanders and P.D. Blankenship, *Mycopathol*, 105, 117 (1989).
[60] F.Y. Du and K.Q. Fu, *Food Sci.* 1, 137 (2008).
[61] J.K. Khalil and M.I.D. Chughtai, *Qual. Plant Foods Hum. Nutr.* 33, 63 (1983).
[62] M.J. Hinds, *Food Chem.* 53, 7 (1994).
[63] N.R. Grosso, V. Nepote and C. Guzmán, *J. Agric. Food Chem.* 48, 806 (2000).
[64] M. Ozcan and S. Seven, *Grasas y Aceites*. 54, 12 (2003).
[65] A.C. Rodrigues, G.L. Ströher, A.R. Freitas, J.V. Visentainer, C.C. Oliveira and N.E. de Souza, *Food Res. Int.* 44, 187 (2011).

In: Peanuts: Bioactivities and Allergies
Editors: I.L. Expósito, A.B. Blázquez ISBN: 978-1-62417-387-5


*Chapter 2*

# PHYSICO-MECHANICAL, AERODYNAMIC, THERMODYNAMIC AND THERMAL PROPERTIES OF PEANUT PODS, KERNELS AND SHELLS

***V. S. P. Bitra****
Associate Professor, College of Agricultural Engineering, Sangareddy, Medak Dist., A.P., India

## ABSTRACT

The peanut (*Arachis hypogaea* Linnaeus) is a major oilseed crop of global significance for processing and consumption. The dimensional, frictional, aerodynamic, and airflow resistance characteristics; water adsorption modeling and moisture adsorption isotherm modelings; and and thermal properties of peanut pods, kernels, and shells were studied for the *cultivar* Tirupathi-2. These properties were studied for a range of moisture contents, and the effect of moisture variation on all the physico-mechanical, aerodynamic, thermodynamic and thermal properties are discussed in detail with the help of graphical representations.

* E-mail address: Prasad_bvss@yahoo.co.in.

## INTRODUCTION

The peanut (*Arachis hypogea* Linnaeus) is a major oilseed crop of global significance for processing and consumption. The peanut is the largest oilseed crop worldwide and is mainly grown during the fall season, in warm climates, and under a wide range of environmental conditions. It is a valuable cash crop for millions of small-scale farmers in the semi-arid tropics. Peanut oil accounts for 55% of the total oil produced from all the oilseeds. The oil content of the seed varies from 44 to 55% depending upon the varieties and agronomic conditions [1].

The application of engineering principles to a biological system is frequently hindered by an inadequacy of data on the basic engineering properties associated with the biological materials involved in the system being considered [2]. Engineering properties are associated with parameters of great importance when analyzing the behavior of grains during their handling, drying, processing, and storage, as well as when designing associated machinery, controls, etc. [3]. These properties are moisture dependent and can affect the adjustments and performance of the processing equipment. A range of moisture content usually exist within which optimum performance is achieved; therefore, the effect of moisture content on the thermal properties of peanut pods, kernels, and shells are of important consideration for post-harvest handling, and food and agricultural processing operations.

The knowledge of physical and thermal properties of these biomaterials is of great importance in the design of handling and processing equipment. The physical properties, such as shape, size, volume, surface area, density, porosity, color and appearance, are important in the design of the machine or the analysis of the product in handling the material. Knowledge of the density and specific gravity of these agricultural products is needed in calculating thermal diffusivity in heat transfer problems, determining the Reynolds number in pneumatic and hydraulic handling of the material, separating the product from undesirable matter, and predicting the physical structure and chemical composition. The static and dynamic angles of repose are imperative in the design of equipment for solid flow and storage structures for the materials [3]. Knowledge of dimensions, aerodynamic properties (terminal velocity and drag coefficient), static and dynamic coefficients associated with the friction of agricultural products is needed in the cleaning, grading, pneumatic and hydraulic handling of the material, separating the product from undesirable matter, predicting a physical structure and constructing the storage structures.

Airflow resistance of various grains and seeds under various conditions has been determined for designing grain aeration and drying systems. To select a fan to adequately deliver air through a grain bed, it is necessary to know how much resistance to overcome. Generally, the static pressure drop across a grain bed is related to the airflow rate, grain bulk density and particle size, as well as surface and shape characteristics of the grain. The stored grain should be aerated in order to maintain a uniform temperature throughout the grain mass to prevent moisture migration, and to prevent spoilage due to mold and insect damage [4]. The peanut shells and pods are porous and hygroscopic materials, whereas the kernels are much denser and less hygroscopic. Therefore, the three materials must be treated separately to adequately describe physical and airflow resistance characteristics.

In studying the behavior of food or agricultural commodities during storing and processing, it is important to have information relating to the environmental conditions associated with the material properties. Equilibrium moisture content (EMC) is one such property and is defined as the moisture content of a hygroscopic material in equilibrium with a particular environment. Equilibrium relative humidity (ERH) is the relative humidity at which the product attains the equilibrium moisture content. A plot of EMCs at a given temperature versus the ERH of the material is called the sorption isotherm. It is important to determine the equilibrium moisture content of agricultural crops for drying, storing, mixing and packaging due to the fact that all agricultural crops are generally hygroscopic. All the agricultural materials have different EMCs at similar conditions because they have different physical and chemical structures. One of the important properties of a material related to drying is its moisture desorption isotherm. Moisture sorption characteristics of food products and their constituents are important for understanding moisture-related degradations such as browning reactions, mold growth, etc. The data on latent heat of vaporization and isosteric heat is needed for determining the efficiency of drying systems. A sorption isotherm depicts the equilibrium relationship between moisture content and water activity at a specific temperature. Knowledge of thermal properties for any food materials is essential for modeling, optimizing and designing the process and equipment for all the operations based on thermal treatments such as dehydrating, blanching, cooking, heating, cooling, steaming, frying, roasting, freezing, etc. [5, 6]. Since new varieties of grains are added to the existing collection faster rates from time to time, there is an urgent need to evaluate and update these properties so that a precise equipment design is possible. An attempt has been made to compile the physico-mechanical, aerodynamic, thermodynamic and thermal properties of peanut pods, kernels, and shells in this chapter for a single *cultivar* Thirupathi-2.

## DIMENSIONAL CHARACTERISTICS

The physical properties, such as length, width, thickness, sphericity, 1,000-grain weight, and shape factor, of peanut pods, kernels and shells of *cultivar* Thirupathi-2 were studied at different moisture contents [7]. The effect of variations in the moisture content (m. c.) for peanut pods, kernels and shells of randomly selected samples gave uneven trends. The uneven trends were due to the random selection of raw materials from bulk at various moisture contents. Hence, for further dimensional characterization, the marked samples were considered for physical measurements. The effect of moisture content variations on physical dimensions, such as average length ($a$), width ($b$) and thickness ($c$), for marked samples of peanut pods, kernels and shells are shown in Figs. 1, 2 and 3, respectively. It was observed that these linear dimensions increased with an increase in moisture content for these materials. A similar trend was observed by earlier researchers [8]. It was observed that the average length, width and thickness of peanut pods increased from 22.72 to 23.47 mm, 11.40 to 11.92 mm and 11.16 to 11.42 mm, respectively, when the moisture content increased from 5.81 to 28.71% (d.b.). In this moisture content range, the average length, width and thickness of pods increased by 3.31, 4.84 and 2.35%, respectively. For the kernels, the average length, width and thickness increased from 10.50 to 10.99 mm, 7.30 to 7.76 mm and 7.03 to 7.23 mm, respectively, when the moisture content increased from 7.62 to 23.67% (d.b.). In this moisture range, the length, width and thickness of kernels increased by 4.73, 6.36 and 2.76%, respectively. Similarly, for peanut shells, these dimensions increased from 22.37 to 23.49 mm, 11.60 to 11.89 mm and 5.59 to 5.92 mm, respectively, when the moisture content increased from 7.71 to 30.03% (d.b.). With this moisture increase, the average length, width and thickness of peanut shells increased by 5.01, 2.57 and 5.83%, respectively. The following general expressions were derived to describe the relationship among the dimensions of pods, kernels and shells.

Pods at 5.81% m.c. (d.b.): $a = 1.993\ b = 2.0364\ c$ (1)

Kernels at 9.16% m.c. (d.b.): $a = 1.4367\ b = 1.4932\ c$ (2)

Shells at 30.03% m.c. (d.b.): $a = 1.9753\ b = 3.9713\ c$ (3)

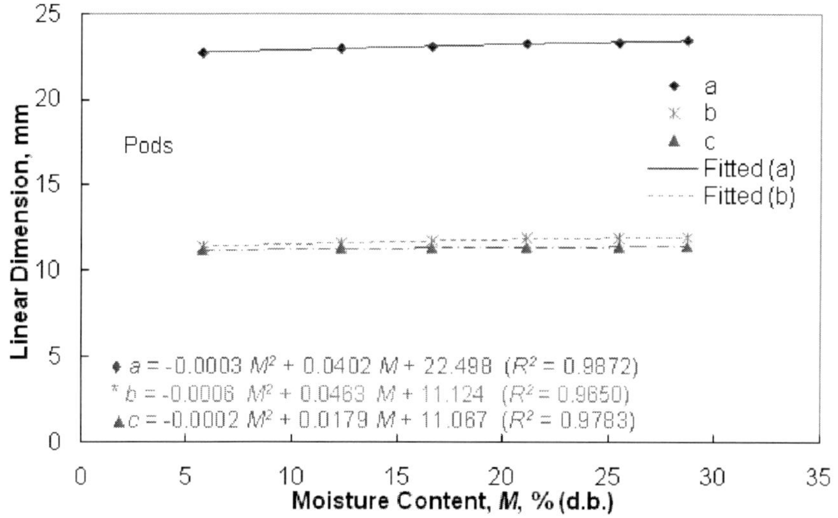


Figure 1. Variation in average dimensions of pods at different moisture contents.

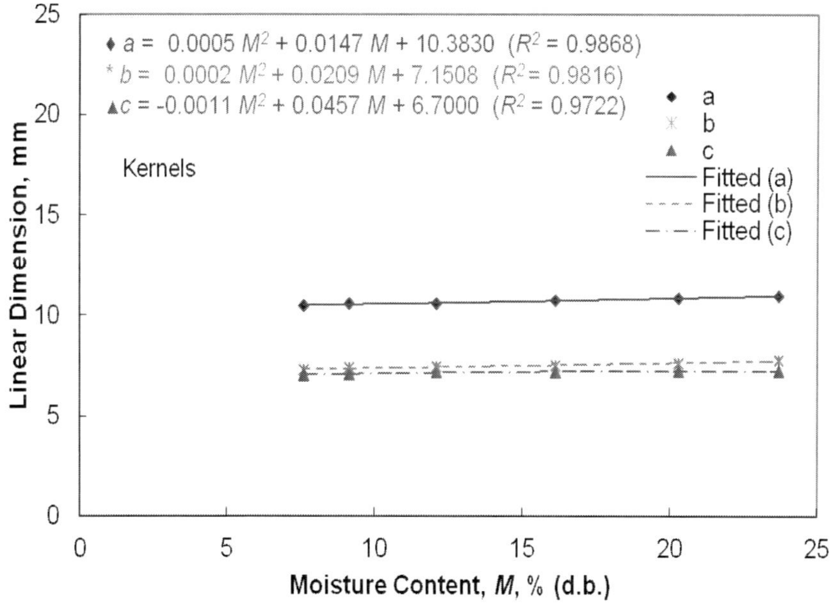


Figure 2. Variation in average dimensions of kernels at different moisture contents.

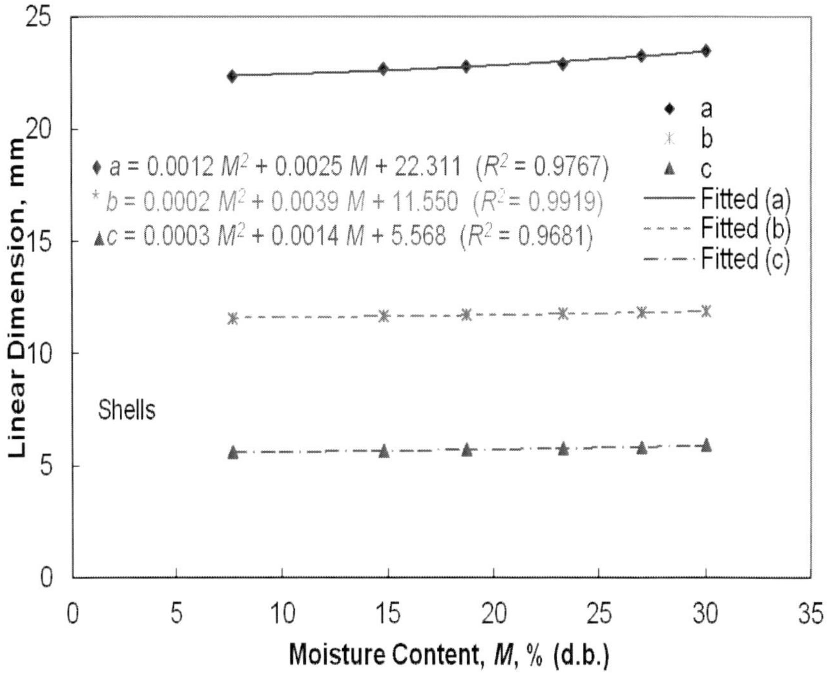


Figure 3. Variation in average dimensions of shells at different moisture contents.

The normal distribution or frequency distribution of the physical dimensions, such as length, width and thickness, of peanut pods, kernels and shells at their respective moisture contents is shown in Figs. 4, 5 and 6, respectively.

It was observed that 82% of the peanut pods at 5.81% moisture content (d.b.) had a length ranging from 19.75 to 24.25 mm. The width and thickness of peanut pods followed a more-or-less similar, normal distribution trend. About 85% of pods had a width ranging from 10.75 to 12.25 mm and about 92% of pods had a thickness ranging from 10.75 to 11.75 mm.

In cases when peanut kernels had a 9.16% moisture content (d.b.), about 76% of them had a length ranging from 9.75 to 11.25 mm, about 76% of them had a width ranging from 6.75 to 8.25 mm and about 82% of them had a thickness ranging from 6.25 to 7.75 mm. Similarly, in cases of peanut shells at 30.03% moisture content (d.b.), about 72% of them had a length ranging from 22.5 to 24.5 mm, whereas about 76% had a width ranging from 11.25 to 12.25 mm and about 69% had a thickness ranging from 6.25 to 7.75 mm. These results indicated that more than two thirds of these materials had dimensions nearer to the central value of the distribution. In all these cases, peak frequencies associated with the lengths of peanut pods, kernels and shells were 40, 36 and 36% with mean dimensions of 22.75, 10.75 and 22.5 mm, respectively.

In cases associated with the width of peanut pods, kernels and shells, the peak frequencies were 32, 32 and 36% with mean dimensions of 10.75, 7.25 and 11.75 mm, respectively.

Similarly, in cases associated with the thickness of peanut pods, kernels and shells, the peak frequencies were 40, 36 and 24% with mean dimensions of 10.75, 6.75 and 7.75 mm, respectively. These results indicated that more than 25% of these materials had lengths, widths and thicknesses approximately equal to the mean dimensions.

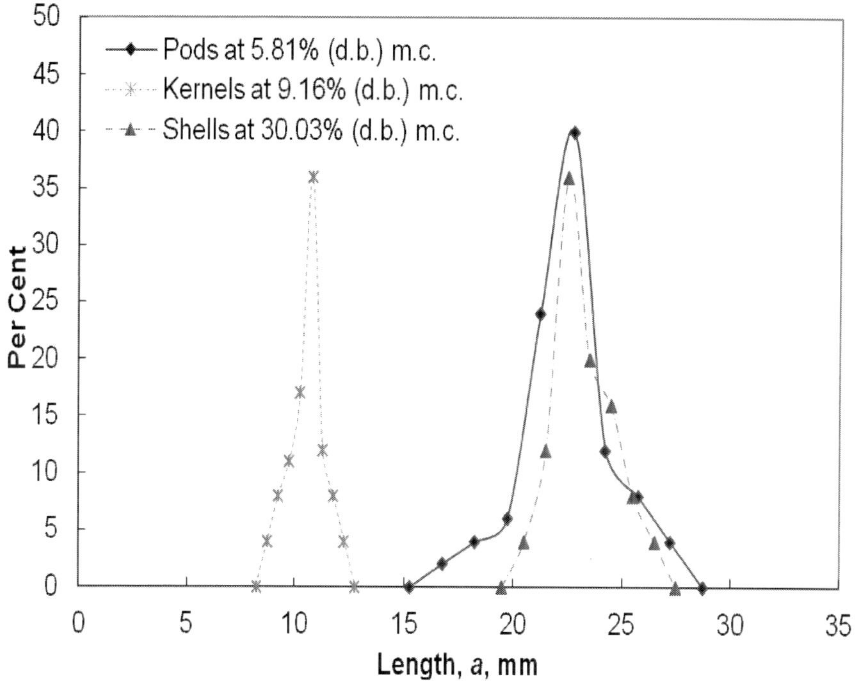


Figure 4. Normal distribution of length for pods, kernels and shells.

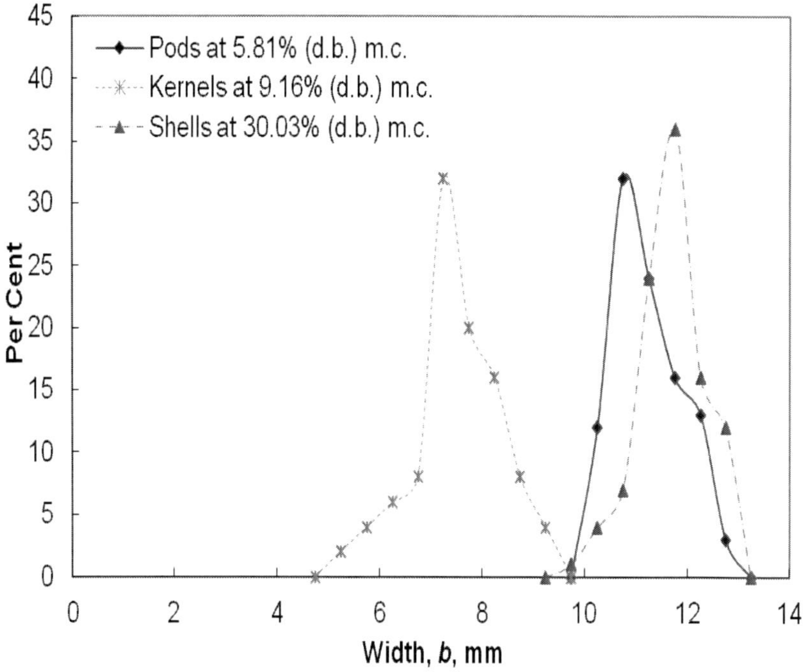


Figure 5. Normal distribution of width for pods, kernels and shells.

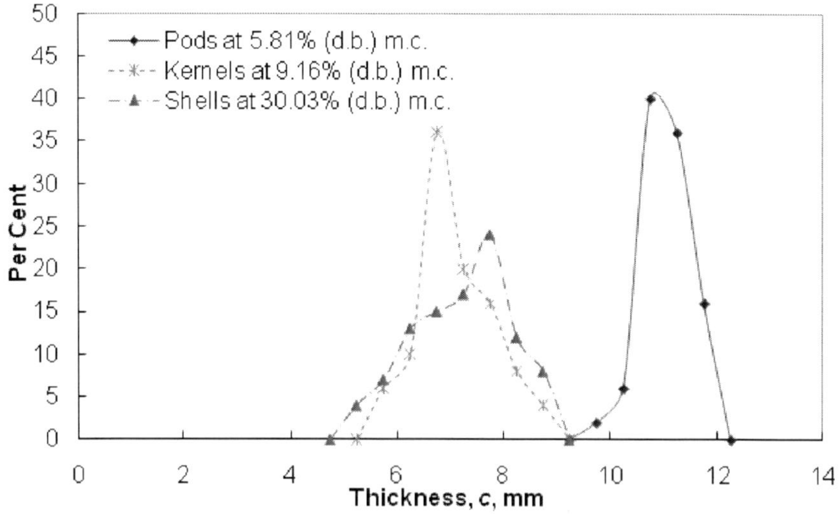


Figure 6. Normal distribution of thickness for pods, kernels and shells.

The effect of variation associated with the moisture content of peanut pods, kernels and shells on the average sphericity is shown in Figure 7. It was observed that the sphericity followed a linear trend with an increase in moisture content. The average sphericity of peanut pods, kernels and shells was found to be 0.64, 0.78 and 0.50, respectively.

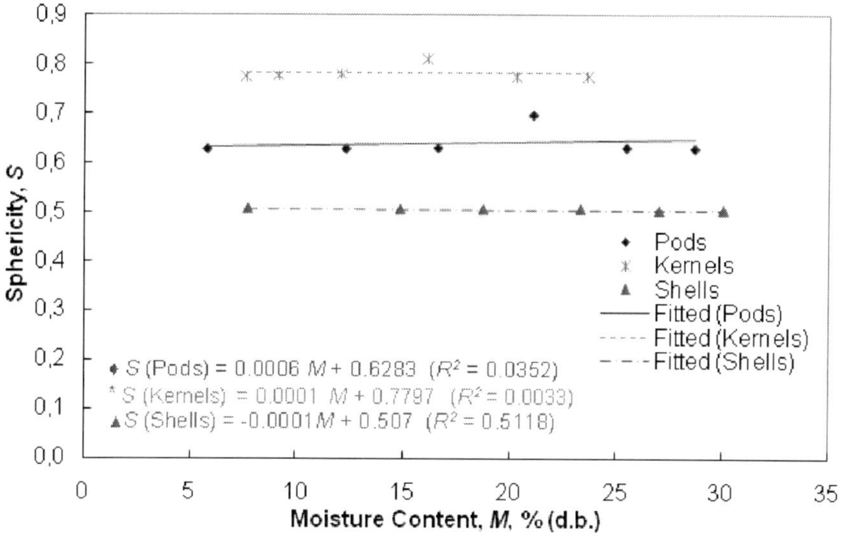


Figure 7. Variation in sphericity of pods, kernels and shells at different moisture contents.

The variation in the 1,000-grain weight of peanut pods, kernels and shells with an increase in moisture content is shown in Figure 8. It was found that the 1,000-grain weight increased with an increase in moisture content. This was due to moisture added to the materials, resulting in a corresponding increase in weight. This result is in accordance with previous research [9, 10, 11]. It was observed that the 1,000-grain weight of pods increased from 649.40 to 790.84 g when the moisture content increased from 6.49 to 31.05% (d.b.). In cases associated with kernels, the weight increased from 269.85 to 334.30 g with an increase in moisture content from 5.98 to 26.25% (d.b.). Similarly, in cases associated with peanut shells, the weight increased from 77.94 to 107.54 g with an increase in moisture content from 7.02 to 36.15% (d.b.). In these ranges of moisture contents for pods, kernels and shells, the 1,000-grain weight increased by 21.78, 23.88 and 37.98%, respectively.

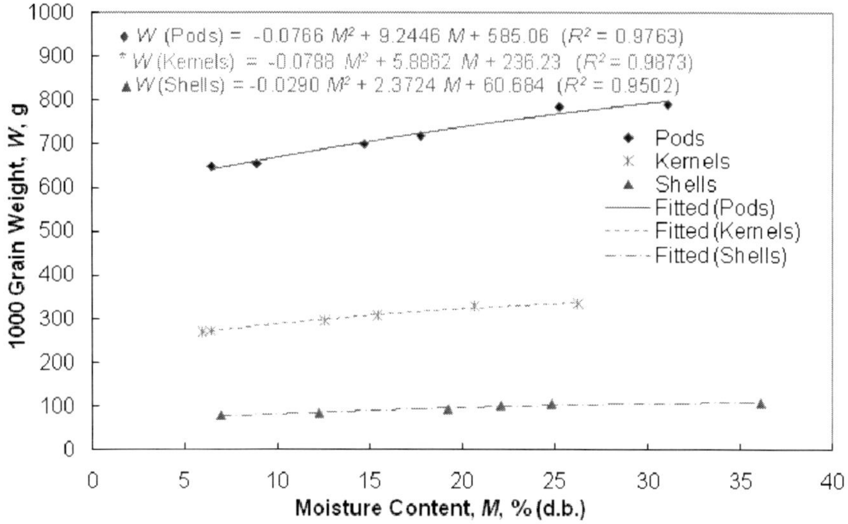


Figure 8. Variation in 1,000-grain weight of pods kernels and shells at different moisture contents.

The shape factor was determined using the sphericity and 1,000-grain weight data. Shape factor, *Z*, was calculated as

$$Z = \frac{\pi D_g^3}{6 D_e^3} \times S \tag{4}$$

where, $D_g$ = geometrical mean diameter, mm = $(abc)^{1/3}$ (5)

$D_e$ = equivalent mean diameter, mm =

$$\left[\frac{6V_{single}}{\pi}\right]^{1/3} \tag{6}$$

$V_{single}$ = volume of single particle, $mm^3$

and *S* = sphericity

In their study, $V_{single}$ was determined using the 1,000-grain weight and particle density.

$$V_{single} = \frac{1000\ grain\ weight \times 10^6}{P.D. \times 1000} \tag{7}$$

where, P.D. = particle density, kg $m^{-3}$.

The particle density at different moisture contents of peanut pods, kernels and shells was determined from the following equations [12]:

$$\text{P.D. (Pod)} = 0.003\ M^2 + 0.5352\ M + 404.83 \tag{8}$$

$$\text{P.D. (Kernels)} = -0.0672\ M^2 + 3.9811\ M + 1281.7 \tag{9}$$

$$\text{P.D. (Shells)} = -0.0013\ M^2 + 0.1799\ M + 271.95 \tag{10}$$

where P.D. = particle density, kg $m^{-3}$

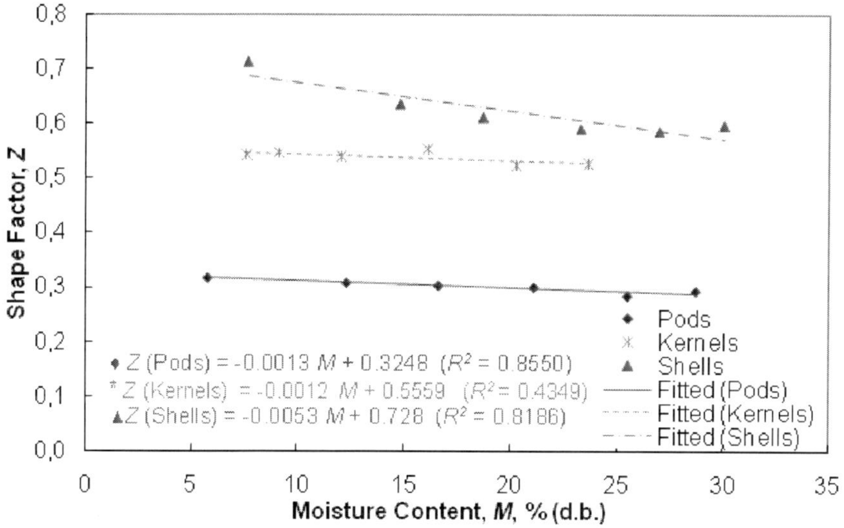


Figure 9. Variation in shape factors of pods, kernels and shells at different moisture contents.

The variation in the shape factor, $Z$, of peanut pods, kernels and shells at different moisture contents is shown in Figure 9. It was observed that the shape factor followed a linear trend with an increase in moisture content. The shape factor of peanut pods, kernels and shells was found to be 0.30, 0.54 and 0.62, respectively. A very high shape factor for shells was found in this study.

This was due to the consideration of the shells' boat-like shape when indirectly determining the $D_e$ values from the 1,000-grain weight and particle density (discussed later) data.

## Bulk Density, Particle Density, and Porosity

The physical properties, such as bulk density, particle density and porosity, for peanut pods, kernels and shells of the same *cultivar* at different moisture contents were studied [12].

The effect of variation associated with the moisture content of peanut pods, kernels and shells on bulk density is shown in Figure 10. It was observed that the bulk density decreased with increases in moisture content for all these materials. The decreasing trend might be due to an increase in the mass and bulk volume of these materials, but the bulk expansion might be higher than the corresponding increase in weight. A similar trend for AK-12-24 and SB-11 varieties of peanut kernels was also observed [8]. The same trend was also observed by several researchers for different grains [13-18]. It was observed that the bulk density decreased from 215.50 to 207.33 kg m$^{-3}$, 713.67 to 692.00 kg m$^{-3}$ and 67.93 to 62.23 kg m$^{-3}$ by increasing the moisture content of peanut pods, kernels and shells from 5.26 to 45.05% (d.b.), 3.64 to 32.05% (d.b.) and 5.69 to 64.89% (d.b.), respectively. When the moisture content of these materials was increased in the range mentioned, the bulk density decreased by 3.80, 3.04 and 8.39% for pods, kernels and shells, respectively.

The particle density was indirectly determined by measuring the porosity with different sand sizes. The cumulative mass fraction of the seashore sand was found with respect to the average particle diameter to characterize the sand used in experimentation. It maintained the general reported trend [19]. The bulk density of various sand sizes was determined. The bulk densities of different average sand sizes of 115, 215 and 327 μm, used in the experiments, were found to be 1,523, 1,525 and 1,540 kg m$^{-3}$.

By using different sand sizes of 115, 215 and 327 μm, the effect of variation associated with the moisture content of peanut pods, kernels and shells on particle density was evaluated [12]. The particle density was observed to be increasing with increases in the moisture content of these materials for all the sand sizes used. The particle density ranged from 400 to 423 kg m$^{-3}$ and from 426 to 441 kg m$^{-3}$ for 5.26 and 45.05% moisture content (d.b.) of pods, from 1,290 to 1,299 kg m$^{-3}$ and from 1,335 to 1,350 kg m$^{-3}$ for 3.64 and 32.05% moisture content (d.b.) of kernels and from 270 to 275 kg m$^{-3}$

and 275 to 281 kg m$^{-3}$ for 5.68 and 64.89% moisture content (d.b.) of shells, respectively, for these sands.

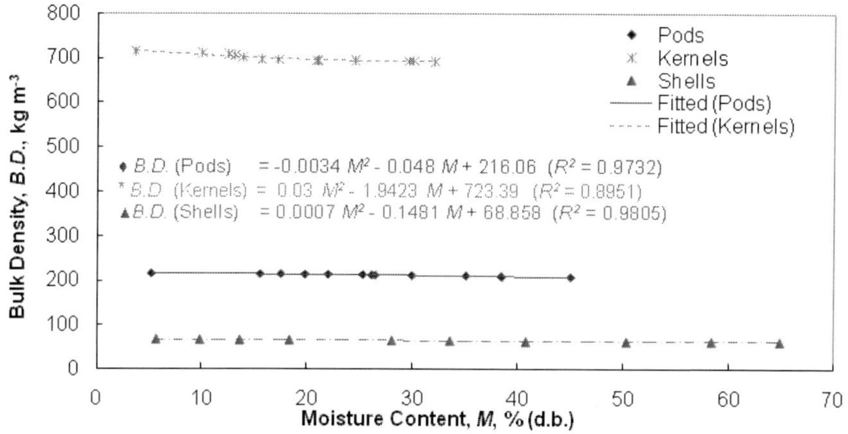


Figure 10. Variation in bulk densities of pods, kernels and shells at different moisture contents.

The variance analysis was studied for the variation of particle density of these materials at each m. c. for the three sizes of sands. It was observed that the variation was insignificant at 5% and 1% levels of significance at all the moisture contents for these materials when different sand sizes were used. As the variation was insignificant, the averaged values of particle densities for each m. c. were calculated. The variation of average particle densities for pods, kernels and shells as a function of moisture content is shown in Figure 11. It was observed that the particle density increased from 409 to 433 kg m$^{-3}$, 1,294 to 1,343 kg m$^{-3}$ and 273 to 278 kg m$^{-3}$ by increasing the moisture content of the peanut pods, kernels and shells from 5.26 to 45.05% (d.b.), 3.64 to 32.05% (d.b.) and 5.69 to 64.89% (d.b.), respectively. In a review of the literature, it was observed that the particle density of most of the grains decreased with increases in moisture content [3, 8, 13, 16]. Contrary to this, an increasing trend of particle density with increases in the moisture content of the rape oilseed was reported by other researchers [15, 17]. The same trend was observed in this study for pods, kernels and shells. This might be due to a swelling of the materials with increases in the moisture content and the corresponding decreases in bulk density. When the moisture content of these materials was increased in the range mentioned, the particle density increased by 3.78, 5.10 and 1.86% for pods, kernels and shells, respectively.

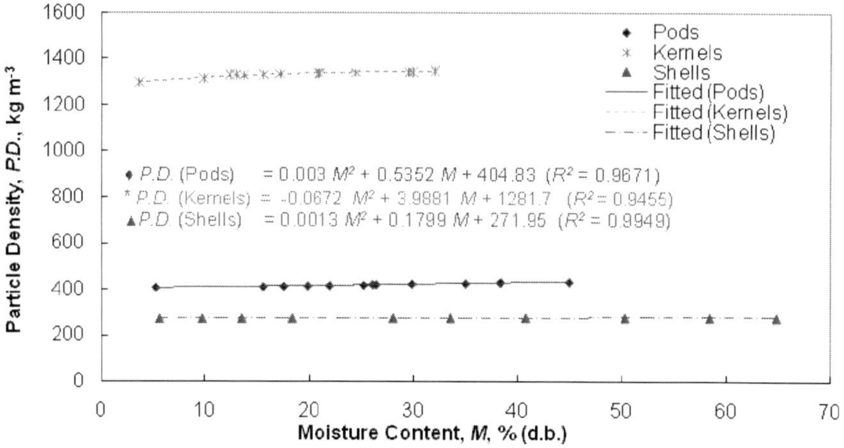


Figure 11. Variation in particle density of pods, kernels and shells at different moisture contents.

By using three sand sizes, porosity variations of peanut pods, kernels and shells at different moisture contents were studied. At each sand size, it was observed that the porosity increased with increases in moisture content for these materials.

The porosity ranged from 45.75 to 49.41% and from 50.70 to 53.24% for 5.26 and 45.05% moisture content (d.b.) of pods, from 45.86 to 46.11% and from 48.17 to 49.58% for 3.64 and 32.05% moisture content (d.b.) of kernels, and from 74.59 to 75.48% and from 77.12 to 77.96% for 5.68 and 64.89% moisture content (d.b.) of shells.

The analysis of variance was studied for the particle density of these materials with different sand sizes. Practically, the particle density was determined from the porosity values.

Thus, the same treatment was applicable to these determinations as well. Hence, the average values of porosity were calculated at each m. c. for these materials, and the variation of average porosity of pods, kernels and shells with increases in moisture content is shown in Figure 12. It was observed that the porosity increased from 47.19 to 52.11%, 45.57 to 48.83% and 75.11 to 77.62% by increasing the moisture content of peanut pods, kernels and shells from 5.26 to 45.05% (d.b.), 3.64 to 32.05% (d.b.) and 5.69 to 64.89% (d.b.), respectively.

A similar trend for AK-12-24 and SB-11 varieties of peanut kernels was also observed [8]. Furthermore, the same trend was also observed by several researchers for different grains [13-15, 17].

This might be due to a swelling of the materials with increases in moisture content. When the moisture content of these materials was increased in the range mentioned, the porosity increased by 10.42, 7.16 and 3.34% for pods, kernels and shells, respectively.

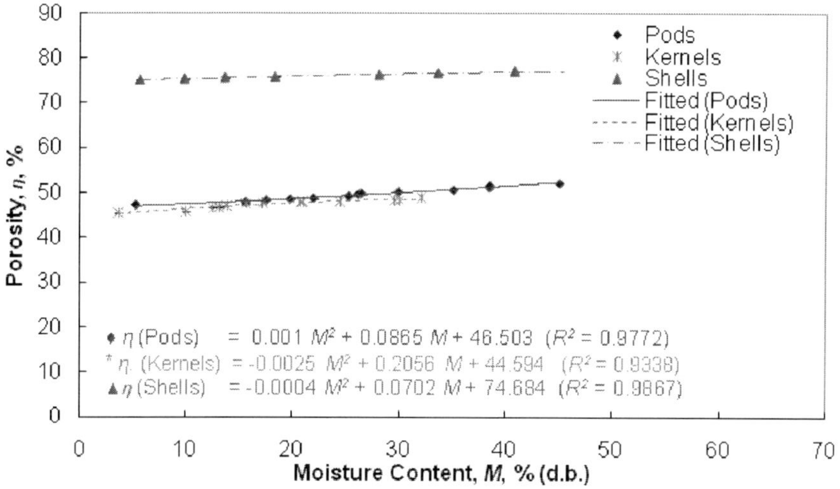


Figure 12. Variation in porosity for peanut pods, kernels and shells at different moisture contents.

## FRICTIONAL CHARACTERISTICS

An experimental set-up for simultaneously determining the sliding angle and angle of repose was fabricated, and these properties were determined at different moisture contents [12]. The sliding angle and angle of repose of peanut pods, kernels and shells of the same *cultivar* at different moisture contents are depicted in Figs. 13, 14 and 15 for pods, kernels and shells, respectively.

The angle of repose of kernels at different moisture contents was also determined with a platform funnel set-up, and the results are shown in Figure 16. It was found that the sliding angle decreased and the angle of repose increased with increases in moisture content for these materials.

The sliding angle decreased from 55.26 to 47.32°, 45.10 to 35.86° and 35.00 to 32.00° by increasing the moisture content of peanut pods, kernels and shells from 5.27 to 45.04% (d.b.), 3.64 to 36.52% (d.b.) and 5.68 to 58.44%

(d.b.), respectively; whereas the angle of repose increased from 30.37 to 34.83°, 24.19 to 29.24° and 32.58 to 36.70° by increasing the moisture content of pods, kernels and shells, respectively, for the above mentioned moisture content. Similar trends associated with the angle of repose for AK-12-24 and SB-11 varieties of peanut kernels were also observed [8].

The trend associated with the angle of repose was in good agreement with past research carried out on various grains [14, 20]. This trend was attributed to increases in cohesive forces between the materials with an increase in moisture content.

When the moisture content of these materials was increased in the range mentioned, the angle of repose increased by 14.68, 20.87 and 12.64% for pods, kernels and shells, respectively; whereas the sliding angle decreased by 14.37, 20.48 and 8.57% for pods, kernels and shells, respectively, in the experimental moisture content range.

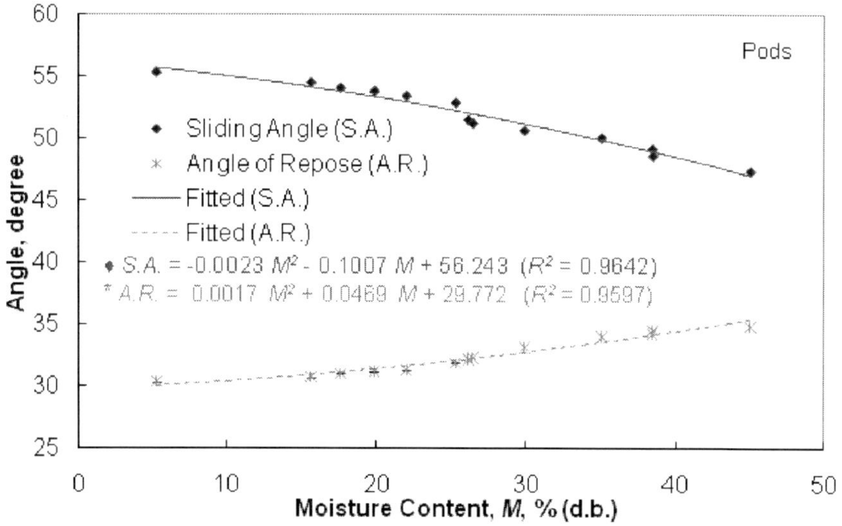


Figure 13. Variation of sliding angles and angles of repose for pods at different moisture contents.

The angle of repose of peanut kernels, studied with the platform-funnel set-up, increased from 32.58 to 36.70° with a similar increase in moisture content from 3.64 to 36.52% (d.b.). These observations were higher than the previous readings taken in the fabricated set-up. In the fabricated angle of repose set-up, the heap of kernels was formed after the material free-flowed over the horizontal platform, which normally induced the sliding of moving materials, resulting in a lesser angle of repose.

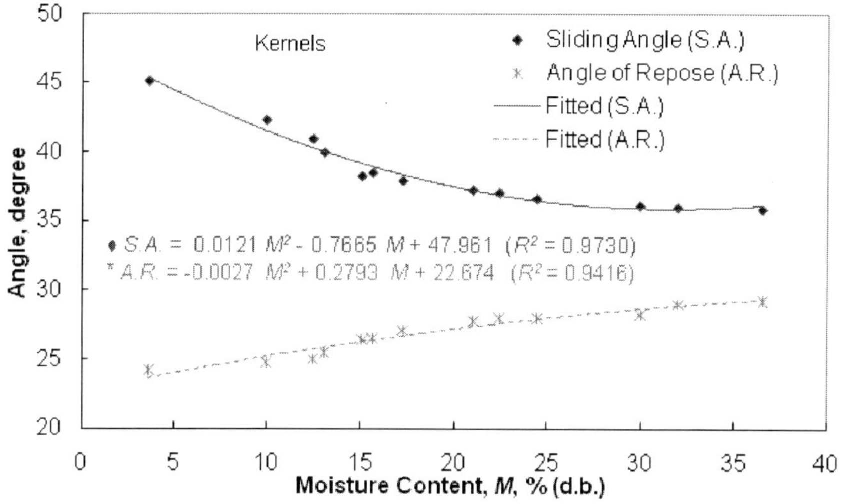


Figure 14. Variation of sliding angles and angles of repose for kernels at different moisture contents.

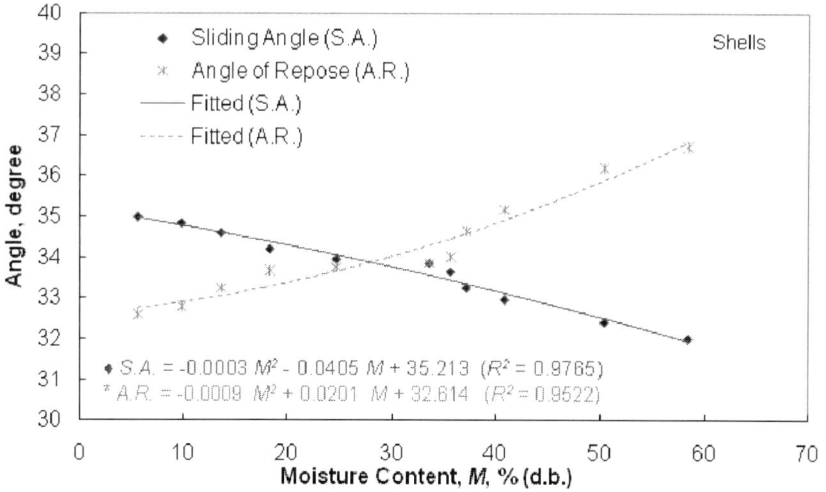


Figure 15. Variation of sliding angles and angles of repose for shells at different moisture contents.

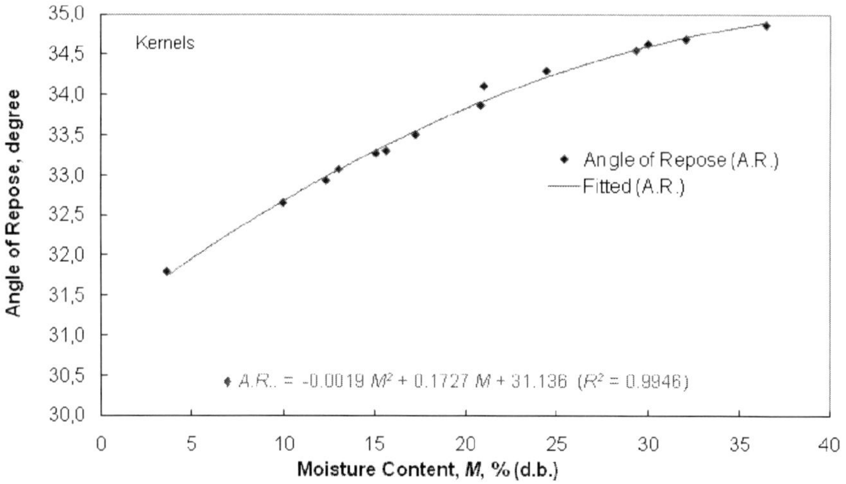


Figure 16. Variation of angles of repose of kernels with moisture using the platform-funnel method.

In contrast, in the platform funnel set-up, the kernels were already in a stationary bed without any disturbance, resulting in a higher angle of repose. The data associated with the sliding angle of repose for these materials can be directly used for designing the hopper bottom and orifices for discharging the granular materials. The angle-of-repose data obtained by the fabricated set-up can be applied for all the cases where the bulk of the materials are in movement while discharging from bins and hoppers.

Experiments were conducted on the same *cultivar* to determine the variation associated with the static coefficients of friction for peanut pods, kernels and shells at different moisture contents over four surfaces, namely a galvanized iron (GI) sheet, mild steel (MS) sheet, plywood with across the grains and plywood with along the grains [7]. The experimental set up consisted of a ring made of a 24-gauge GI sheet (100 mm diameter and 50 mm height) which was bottomless and topless, and an aluminium pan connected by a thin twine over a pulley.

The effect associated with variations of moisture content for peanut pods, kernels and shells on static coefficients of friction over a GI sheet, MS sheet, plywood with across the grains and plywood with along the grains are shown in Figs. 17, 18, 19, and 20, respectively. In all the cases, the static coefficient of friction increased with an increase in moisture content. It was found that the static coefficient of friction increased from 0.5 to 0.57, 0.35 to 0.46 and 0.42 to 0.62 over the GI sheet for peanut pods, kernels and shells with an increase in moisture content from 6.49 to 31.05% (d.b.), 5.98 to 26.25% (d.b.) and 7.02 to 36.15% (d.b.), respectively. For these moisture ranges, the static coefficient of friction increased by 14.0, 31.7 and 46.7% for pods, kernels and shells, respectively.

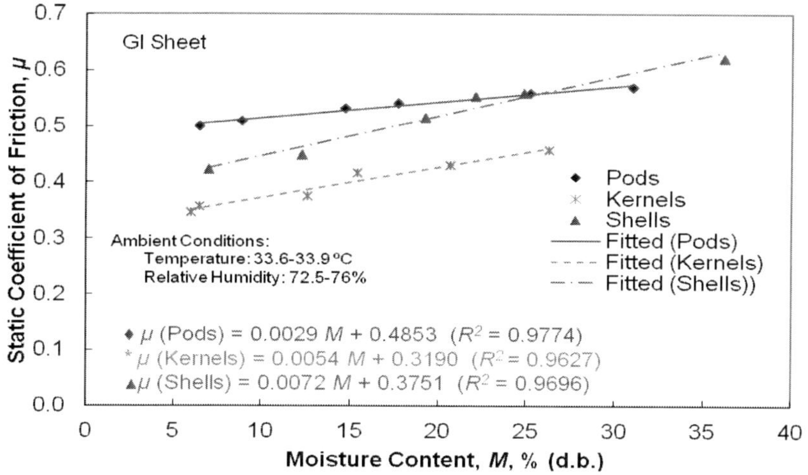


Figure 17. Variation of static coefficients of friction with moisture content on a GI sheet.

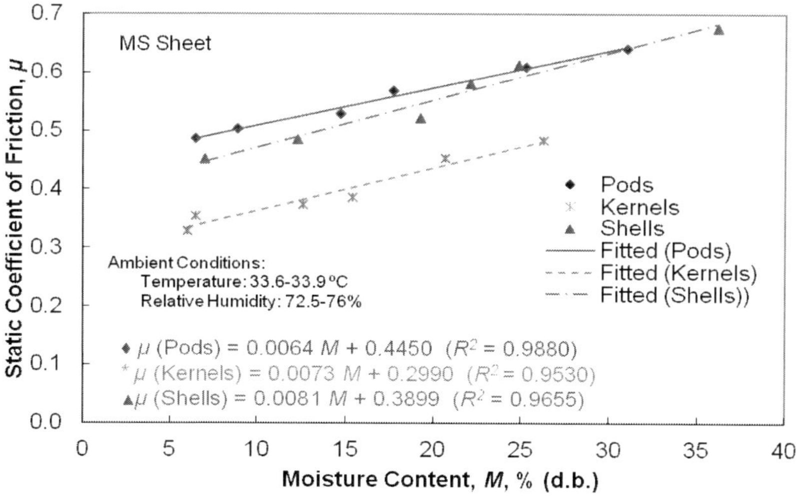


Figure 18. Variation of static coefficients of friction with moisture content on an MS sheet.

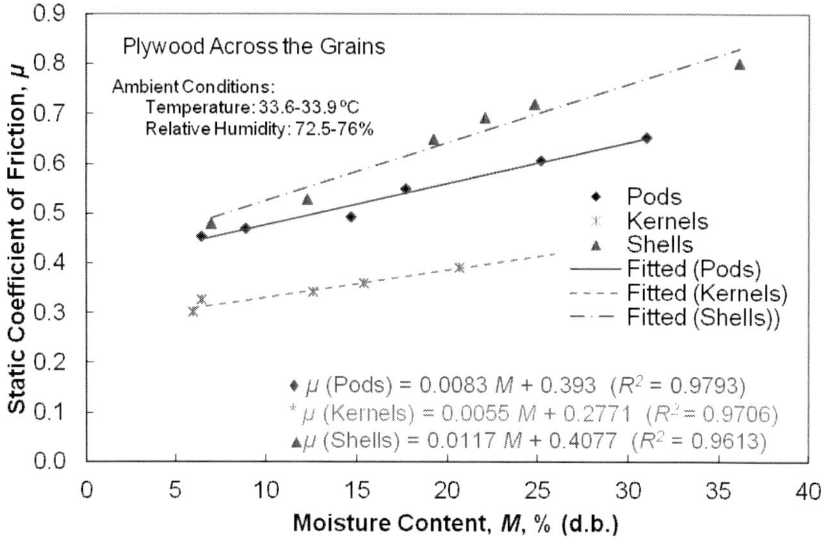


Figure 19. Variation of static coefficients of friction with moisture content on plywood (across).

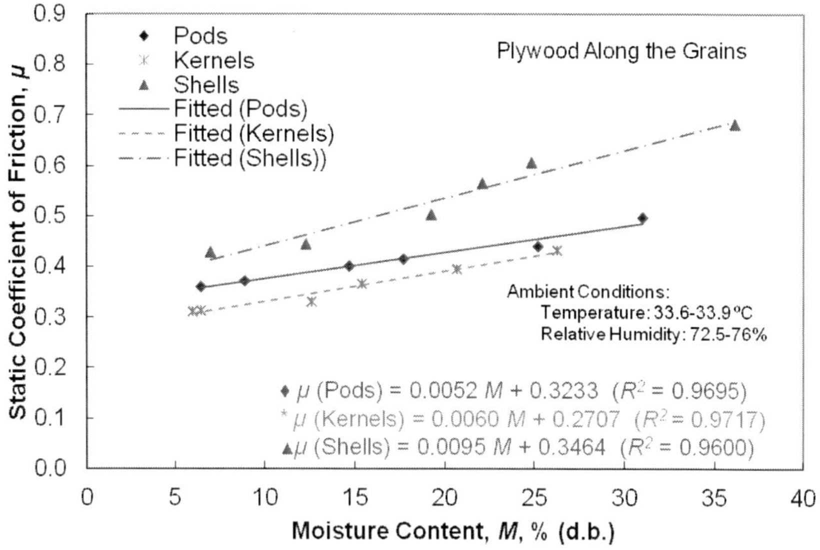


Figure 20. Variation of static coefficients of friction with moisture content on plywood (along).

In the case of the mild steel sheet, the static coefficient of friction increased from 0.49 to 0.64, 0.33 to 0.48 and 0.45 to 0.68 for pods, kernels and shells with an increase in moisture content from 6.49 to 31.05% (d.b.), 5.98 to 26.25% (d.b.) and 7.02 to 36.15% (d.b.), respectively. With these moisture ranges, the coefficient of friction for pods, kernels and shells increased by 31.83, 47.56 and 49.78%, respectively.

Likewise, over plywood with across the grains, the static coefficient of friction for pods, kernels and shells increased from 0.45 to 0.65, 0.30 to 0.42, and 0.48 to 0.80 with an increase in moisture content from 6.49 to 31.05% (d.b.), 5.98 to 26.25% (d.b.) and 7.02 to 36.15% (d.b.), respectively. With these moisture ranges, the static coefficient of friction increased by 43.61, 40.65 and 67.43% for peanut pods, kernels and shells, respectively. Similarly, over plywood with along the grains, the coefficient of friction for peanut pods, kernels and shells increased from 0.36 to 0.50, 0.31 to 0.43 and 0.43 to 0.68 with an increase in moisture content from 6.49 to 31.05% (d.b.), 5.97 to 26.25% (d.b.) and 7.02 to 36.15% (d.b.), respectively. With these moisture ranges, the static coefficient of friction increased by 38.16, 38.46 and 59.21%, respectively. It was found that the static coefficient of friction for peanut pods, kernels and shells on GI and MS sheets was more-or-less equal.

In the case of plywood, the static coefficient of friction across the grains was more than that of plywood with along the grains, and the static coefficient of friction for peanut pods, kernels and shells on plywood was more than that on GI and MS sheets.

The static and dynamic coefficients of friction for peanut pods, kernels, and shells were determined [21]. It consisted of a frictionless pulley with a 35 mm diameter fixed to the Shimadzu Food Tester, Japan (Model: EZTest/CE), interfaced with a Vintron PIII computer. The circular sample holder was fabricated with a 22-gauge galvanized iron (GI) sheet (180 mm diameter and 100 mm height) which was bottomless and topless. A steel cable wire thathad a 2 mm diameter was fixed at the bottom of the sample holder via an eye bolt and passed around the pulley.

The axis of the eye bolt was 10 mm above the test surface. This location is necessary is in order to minimize the rotational coupling with respect to the lower edge of the sample holder. The pulley was supported on a frictionless shaft made of mild steel. The other end of the wire was fixed to the load transducer (capacity: 500 N) of the Food Tester.

The sample holder was placed over different test surfaces such as a GI sheet (24 gauge), mild steel (MS) sheet (18 gauge) and plywood (3 mm thickness) with across the grains and along the grains. The test surfaces could be lifted or lowered using a jockey in order to maintain the horizontal position to the pulling wire. The jockey was made stable by welding 16-gauge MS sheets on the bottom and top. The test surfaces were fixed rigidly to the top plate of the jockey with fasteners. A height gauge was used to confirm the horizontal position of the wire. The sample in the holder was facilitated to load from the top by using a canvas bag containing sand and machined standard iron weights for further surcharged loading.

The angle of internal friction was determined by pulling the sample over the same sample. During this experimentation, the bottom sample was kept stationary in an elongated D-shaped container, fabricated with a 22-gauge GI sheet. This container was fixed to the test surface without any movement. The sample in the bottom container was maintained stationary by using a 22-gauge GI sheet moving along with the top circular sample holder.

The effect associated with the variation of moisture contents for peanut pods, kernels, and shells on the static coefficient of friction over GI sheets, MS sheets, plywood with across the grains and plywood with along the grains are shown in Figs. 21, 22, and 23, respectively. In all the cases, the static coefficient of friction increased with an increase in moisture content. It was found that the static coefficient of friction increased from 0.22 to 0.23, 0.06 to 0.08, and 0.10 to 0.23, over GI sheets for peanut pods, kernels, and shells with an increase in moisture content from 12.20 to 24.16% (d.b.), 10.35 to 23.31% (d.b.), and 7.66 to 33.54% (d.b.), respectively. For these moisture ranges, the static coefficient of friction increased by 4.0, 33.33, and 129.70% for peanut pods, kernels, and shells, respectively. These results were in accordance with the observations of past research [22] for unshelled peanuts over sheet metal. In the case of the MS sheet, the static coefficient of friction increased from 0.21 to 0.24, 0.08 to 0.12, and 0.13 to 0.29 for peanut pods, kernels, and shells with an increase in moisture content from 12.20 to 24.16% (d.b.), 10.35 to 23.31% (d.b.), and 7.66 to 33.54% (d.b.), respectively. With these moisture ranges, the coefficient of friction for peanut pods, kernels, and shells increased by 14.22, 56.59, and 128.00%, respectively.

The values of static coefficients of friction for peanut pods over the MS sheet were much less than the earlier obtained results [23]. Likewise, over plywood with across the grains, the static coefficient of friction for peanut pods, kernels, and shells increased from 0.22 to 0.33, 0.10 to 0.16, and 0.22 to 0.27 with an increase in moisture content from 12.20 to 24.16% (d.b.), 10.35 to 23.31% (d.b.), and 7.66 to 33.54% (d.b.), respectively. With these moisture ranges, the static coefficient of friction increased by 52.29, 64.84, and 19.20% for peanut pods, kernels, and shells, respectively. These results were in accordance with the results obtained earlier [24] for peanut kernels over plywood. Similarly, over plywood with along the grains, the coefficient of static friction for peanut pods, kernels, and shells increased from 0.14 to 0.26, 0.05 to 0.11, and 0.09 to 0.21 with an increase in moisture content from 12.20 to 24.16% (d.b.), 10.35 to 23.31% (d.b.), 7.66 to 33.54% (d.b.), respectively. With these moisture ranges, the static coefficient of friction increased by 80.70, 117.14, 120.00%, respectively.

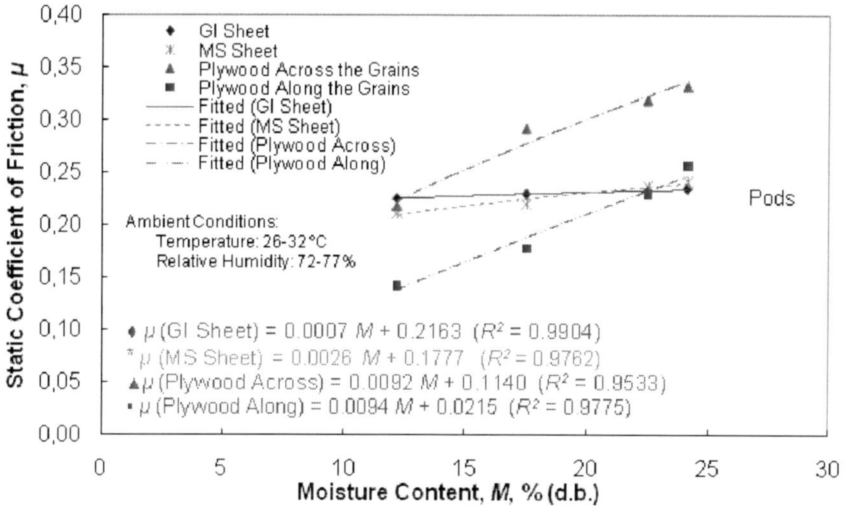


Figure 21. Variation in static coefficients of friction for pods at different moisture contents.

It was found that the static coefficient of friction for peanut pods, kernels, and shells on MS sheets was more than that on the GI sheet. In the case of plywood, the static coefficient of friction across the grains was more than that of the plywood with along the grains.

The static coefficient of friction of peanut pods, kernels, and shells on plywood with across the grains was more, and on plywood with along the grains, it was less than on GI and MS sheets. In general, the values of static coefficients of friction were less than the results obtained [7] for peanut pods, kernels and shells of the same variety. The reason might be due to changes in the experimental procedure and ambient conditions.

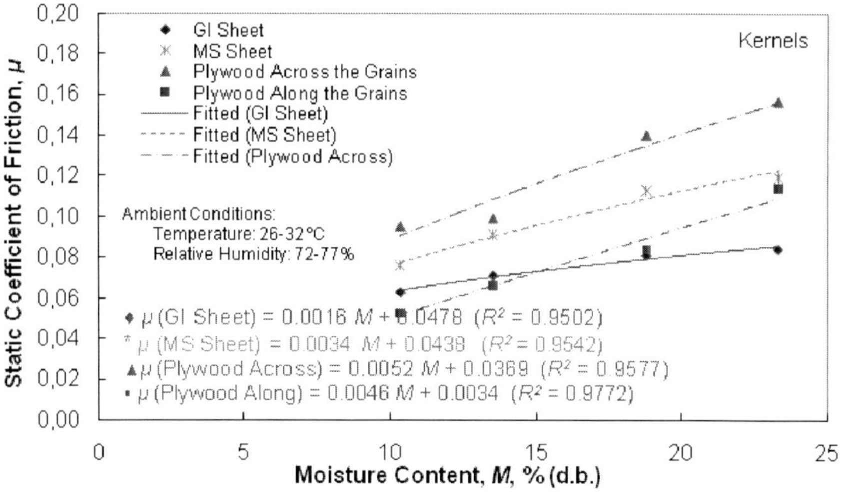


Figure 22. Variation in static coefficients of friction for kernels at different moisture contents.

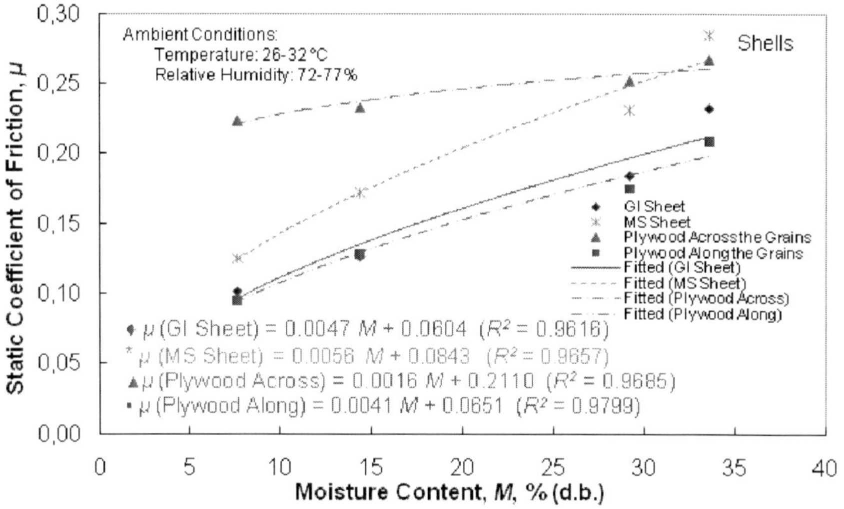


Figure 23. Variation in static coefficients of friction for shells at different moisture contents.

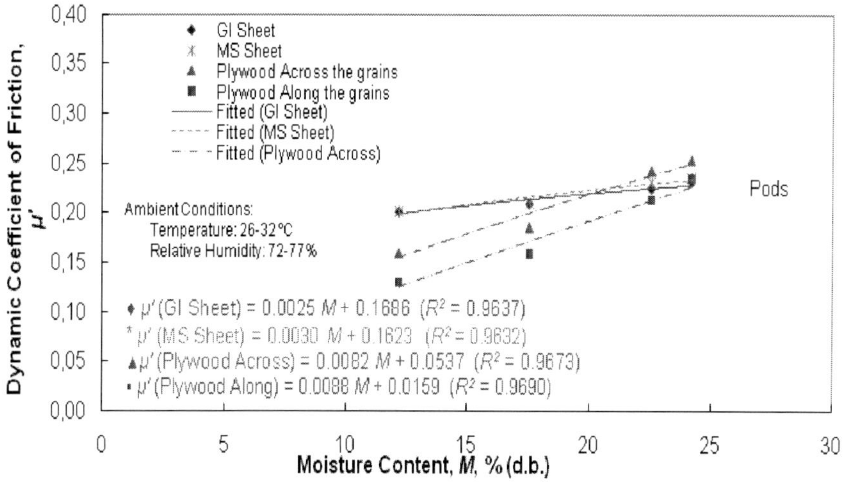


Figure 24. Variation in dynamic coefficients of friction for pods at different moisture contents.

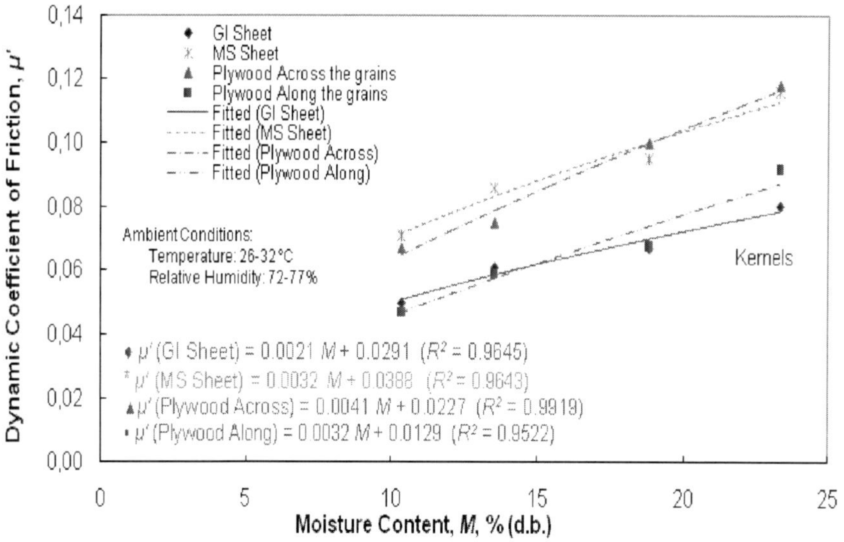


Figure 25. Variation in dynamic coefficients for friction of kernels at different moisture contents.

The effect associated with the variation of moisture content for peanut pods, kernels, and shells on the dynamic coefficient of friction over a GI sheet, MS sheet, plywood with across the grains and plywood with along the grains is shown in Figs. 24, 25, and 26, respectively. In all the cases, dynamic coefficients of friction increased with an increase in moisture content. It was observed that the dynamic coefficient of friction was less than the static coefficient of friction as reported earlier [25]. It was found that the dynamic coefficient of friction increased from 0.20 to 0.23, 0.05 to 0.08, and 0.17 to 0.22 over the GI sheet for peanut pods, kernels, and shells with an increase in moisture content from 12.20 to 24.16% (d.b.), 10.35 to 23.31% (d.b.), and 7.66 to 33.54% (d.b.), respectively. For these moisture ranges, the dynamic coefficient of friction increased by 14.93, 60.00, and 211.59% for peanut pods, kernels, and shells, respectively. These results were in accordance with the data obtained by the earlier researchers [22] for unshelled peanuts over sheet metal.

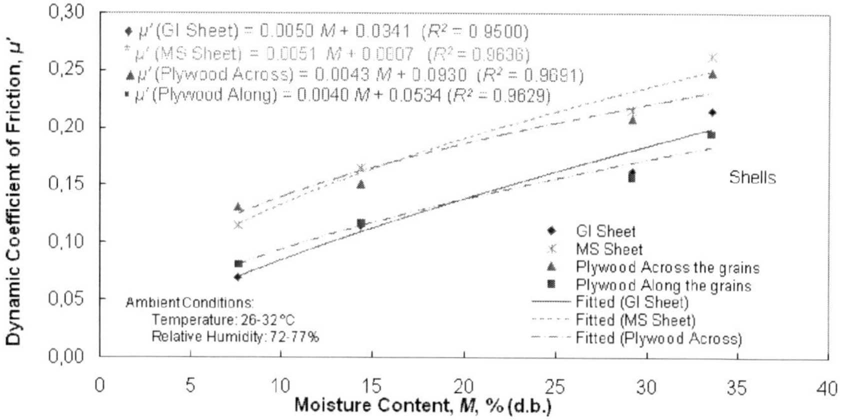


Figure 26. Variation in dynamic coefficients of friction for shells at different moisture contents.

In the case of the MS sheet, the dynamic coefficient of friction increased from 0.20 to 0.24, 0.07 to 0.12, and 0.12 to 0.26 for peanut pods, kernels, and shells with an increase in moisture content from 12.20 to 24.16% (d.b.), 10.35 to 23.31% (d.b.), and 7.66 to 33.54% (d.b.), respectively. With these moisture ranges, the coefficient of friction for peanut pods, kernels, and shells increased by 16.92, 63.38, and 128.69%, respectively.

Likewise, over plywood with across the grains, the dynamic coefficient of friction of peanut pods, kernels, and shells increased from 0.16 to 0.25, 0.07 to 0.12, and 0.13 to 0.25 with an increase in moisture content from 12.20 to 24.16% (d.b.), 10.35 to 23.31% (d.b.), and 7.66 to 33.54% (d.b.), respectively. With these moisture ranges, the dynamic coefficient of friction increased by

58.75, 76.56, 89.31% for peanut pods, kernels, and shells, respectively. Similarly, over plywood with along the grains, the coefficient of dynamic friction for peanut pods, kernels, and shells increased from 0.13 to 0.24, 0.05 to 0.09, and 0.08 to 0.19 with increase in moisture content from 12.20 to 24.16% (d.b.), 10.35 to 23.31% (d.b.), 7.66 to 33.54% (d.b.), respectively. With these moisture ranges, the dynamic coefficient of friction increased by 80.76, 95.74, 140.74%, respectively.

It was found that the dynamic coefficient of friction for peanut pods, kernels, and shells on the MS sheet was more than that on the GI sheet. In the case of plywood, the dynamic coefficient of friction across the grains was more than that of the plywood with along the grains.

The dynamic coefficient of friction for peanut pods, kernels, and shells on plywood with along the grains was less than that on GI and MS sheets.

## Mechanical Properties

Experiments were conducted on mechanical properties to determine the variation of shear force for peanut pods, kernels, and shells at different moisture contents [21].

The effect associated with the variation of moisture content for peanut pods, kernels, and shells on shear force with an increase in normal force is shown in Figs. 27, 28, and 29, respectively. In all the cases, shear force increased with an increase in normal force. It was observed that the shear force decreased with an increase in moisture content of the samples at each normal force. It was found that the shear force increased from 46.17 to 80.60 N, 39.58 to 78.04 N, 31.84 to 75.25 N and 24.16 to 68.31 N with an increase in normal force from 61.244 to 167.155 N for peanut pods at the moisture contents of 12.20, 17.55, 22.53 and 24.16% (d.b.), respectively. For these moisture contents, the shear force increased by 74.55, 97.20, 136.36 and 182.72% for peanut pods with an increase in normal force from 61.244 to 167.155 N.

In the case of peanut kernels, it was found that the shear force increased from 32.09 to 56.60 N, 26.48 to 52.83 N, 24.24 to 54.33 N and 20.96 to 62.93 N with an increase in normal force from 63.695 to 169.606 N for peanut kernels at the moisture contents of 10.35, 13.53, 18.80 and 23.31% (d.b.), respectively. For these moisture contents, the shear force increased by 76.39, 99.53, 124.13 and 152.50% for peanut kernels with an increase in normal force from 63.695 to 169.606 N.

Likewise, for peanut shells, it was observed that the shear force increased from 45.99 to 79.27 N, 32.00 to 72.01 N, 26.60 to 70.33 N and 23.86 to 69.86 N with an increase in normal force from 59.871 to 165.780 N for peanut shells at the moisture contents of 7.66, 14.38, 29.17 and 33.54% (d.b.), respectively. For these moisture contents, the shear force increased by 72.37, 125.03, 164.37 and 192.76% for peanut shells with an increase in normal force from 59.871 to 165.780 N.

The slope of the curves was equivalent to the coefficient of internal friction, (i.e., Tan$\phi_i$, from which the angle of internal friction, $\phi_i$, was found out). The constant in the linear equation was considered as a cohesive force, *C*. The effect associated with the variation of moisture content for peanut pods, kernels, and shells on the angle of internal friction is shown in Figure 30. It was observed that the angle of internal friction, $\phi_i$, increased non-linearly with an increase in moisture content for all the samples. The angle of internal friction was in the order of peanut shells> peanut pods> peanut kernels.

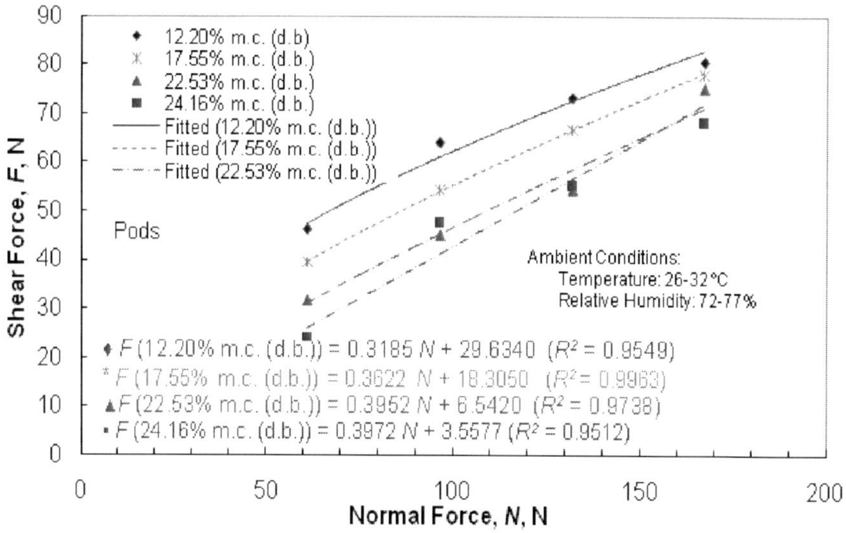


Figure 27. Variation in the shear force of pods at different moisture contents with normal force.

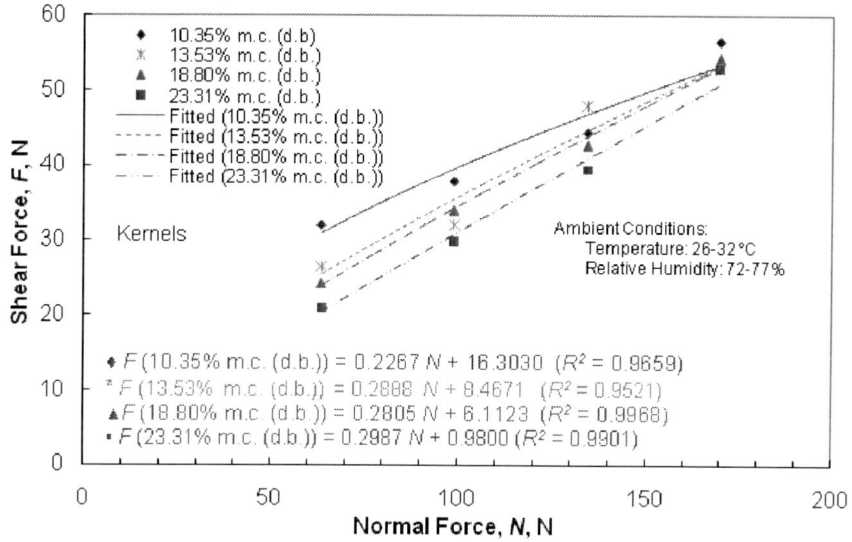


Figure 28. Variation in the shear force of kernels at different moisture contents with normal force.

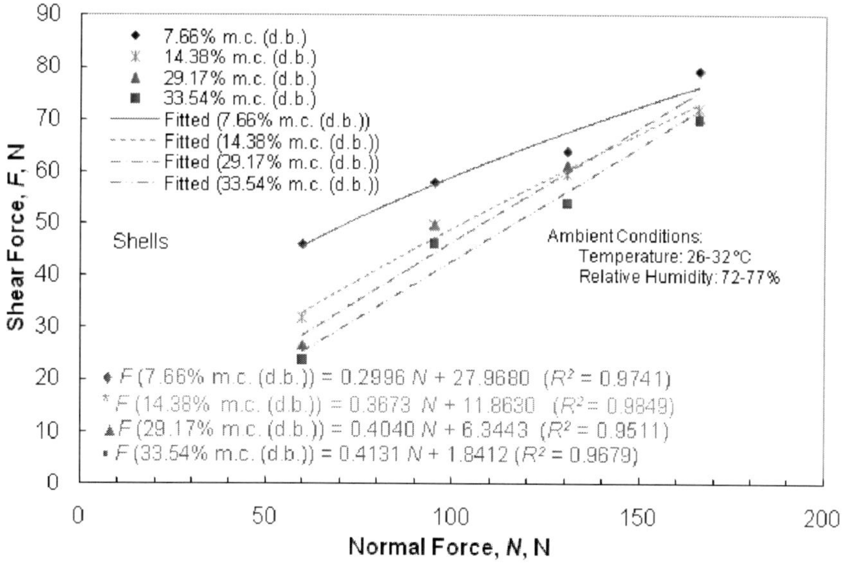


Figure 29. Variation in the shear force of shells at different moisture contents with normal force.

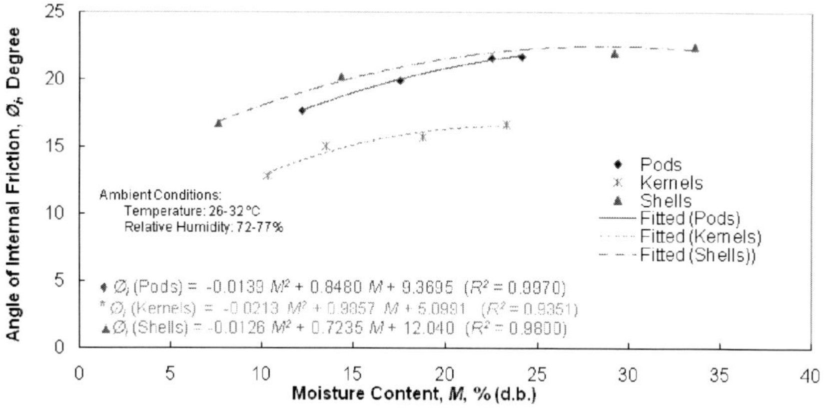


Figure 30. Variation in the angle of internal friction at different moisture contents.

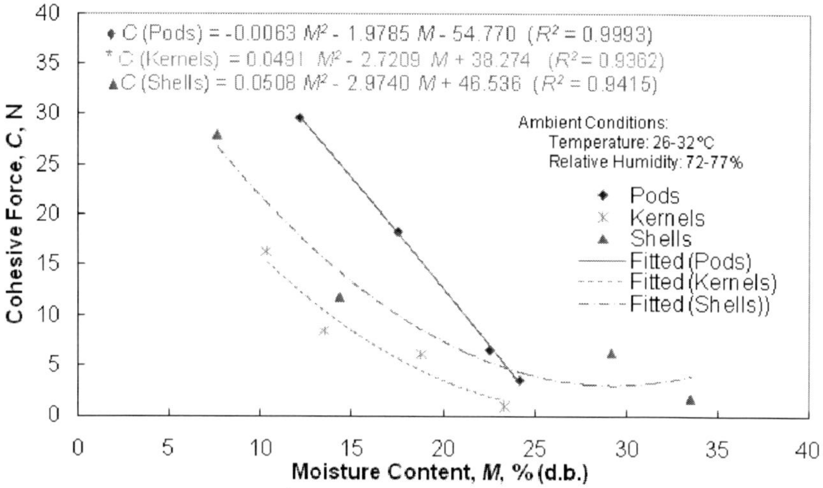


Figure 31. Variation in the cohesive force at different moisture contents.

It was found that the angle of internal friction increased from 17.67 to 21.66°, 12.77 to 16.63°, and 16.68 to 22.45° for peanut pods, kernels, and shells with an increase in moisture content from 12.20 to 24.16% (d.b.), 10.35 to 23.31% (d.b.), and 7.66 to 33.54% (d.b.), respectively. For these moisture ranges, the angle of internal friction increased by 22.63, 30.21, 34.58% for peanut pods, kernels, and shells, respectively.

The cohesive force was taken from the linear equations of the shear force versus normal force curves. The effect associated with the variation of moisture content for peanut pods, kernels, and shells on cohesive force is shown in Figure 31. It was observed that the cohesive force, *C*, decreased non-linearly with an increase in moisture content for all the samples. The cohesive force was in the order of peanut pods> peanut shells> peanut kernels. It was found that the cohesive force decreased from 29.63 to 3.56 N, 16.31 to 0.98 N, and 27.97 to 1.84 N for peanut pods, kernels, and shells with an increase in moisture content from 12.20 to 24.16% (d.b.), 10.35 to 23.31% (d.b.), and 7.66 to 33.54% (d.b.), respectively. For these moisture ranges, the cohesive force decreased by 87.99, 93.99, 93.42% for peanut pods, kernels, and shells, respectively.

## AERODYNAMIC PROPERTIES

Experiments were also conducted [7] on the same *cultivar* to determine the variation of terminal velocities and drag coefficients for peanut pods, kernels and shells at different moisture contents. A wind tunnel with a centrifugal air blower was modified to determine the terminal velocities. The experimental set-up consisted of a 0.25 hp centrifugal blower, plenum chamber, vertical Perspex tube, air flow control ring (25 mm length, 46 mm ID and 50 mm OD) and an inlet pipe. The vertical Perspex tube of 923 mm length, 40 mm ID and 45 mm OD was fixed at the top of the plenum chamber with three screws and a height of 500 mm was marked on the tube. At the top of the vertical tube, the opening was closed up to a 23.5 mm diameter with a Perspex sheet and covered with an air control ring which was graduated around its periphery. The ring opening was closed up to 23 mm with the Perspex sheet. By operating the ring, the air flow rate in the tube was controlled. At the inlet of the blower, a Perspex tube with a length of 300 mm, 84 mm ID and 90 mm OD was fitted to determine the air velocity using a digital vane anemometer placed at one end of the Perspex tube. The velocity of air was recorded at the inlet pipe whenever the marked samples lifted up to the 500 mm mark on the vertical tube, and the terminal velocity was calculated.

The drag coefficient, $C_d$, of peanut pods, kernels and shells at different moisture contents was determined using the following equation:

$$C_d = \frac{2\, m\, g\, (\rho_p - \rho_a)}{v_t^2\, \rho_a\, A\, \rho_p} \qquad (11)$$

where, $m$ =mass of single particle, kg

$g$ =acceleration due to gravity, m $s^{-2}$

$v_t$ =average terminal velocity, m $s^{-1}$

$\rho_a$ =density of air, kg $m^{-3}$ (from psychrometric chart)

$\rho_p$ =particle density of particle, kg $m^{-3}$

$A$ = projected area of particle, $m^2$ =

$$\frac{\pi}{4}D_e^2 \tag{12}$$

and $D_e$ =equivalent mean diameter, m.

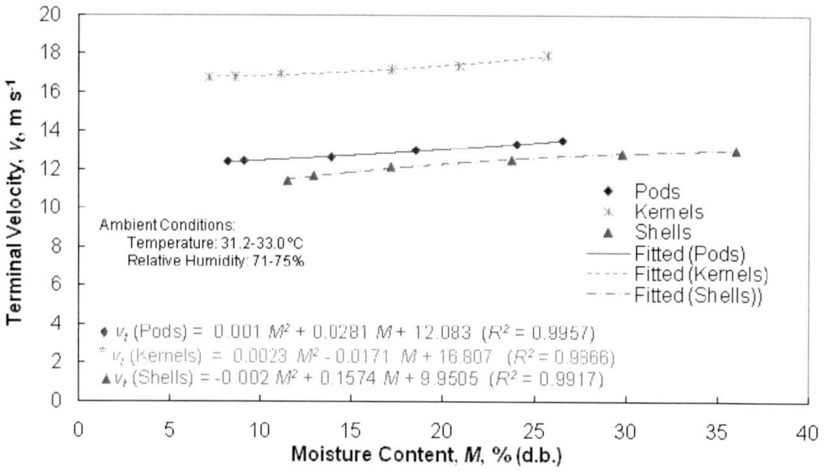


Figure 32. Variation in terminal velocities for pods, kernels, and shells at different moisture contents.

The effect associated with the variation of moisture content on the average terminal velocity of peanut pods, kernels and shells is shown in Figure 32. It was observed that the terminal velocity of these materials increased with an increase in moisture content. These results are in accordance with different agricultural materials studied by various researchers [26-33]. It was observed that the terminal velocity increased from 12.39 to 13.51 m $s^{-1}$ when the moisture content of peanut pods rose from 8.24 to 26.46% (d.b.). In the case of the peanut kernels, it increased from 16.77 to 17.92 m $s^{-1}$ with an increase in moisture content from 7.20 to 25.67% (d.b.). Similarly, in case of peanut shells, with an increase in moisture content from 11.52 to 36.01% (d.b.), the terminal velocity increased from 11.42 to 12.99 m $s^{-1}$. The terminal velocity of kernels was high, followed by peanut pods and shells. The higher values for kernels were due to a higher particle density than pods and shells. A marginal decrease in the values of terminal velocity was observed for shells, compared to peanut pods. The reason might be a marginal difference in the particle density of these materials. Particle density of the same *cultivar* increased from 409 to 433 kg $m^{-3}$, 1,294 to 1,343 kg $m^{-3}$ and 273 to 278 kg $m^{-3}$ by increases in the moisture content of peanut pods, kernels and shells from 5.26 to 45.05% (d.b.), 3.64 to 32.05% (d.b.) and 5.69 to 64.89% (d.b.), respectively [12]. In the moisture range studied, the terminal velocity of peanut pods, kernels and shells increased by 9.04, 6.84 and 13.70%, respectively.

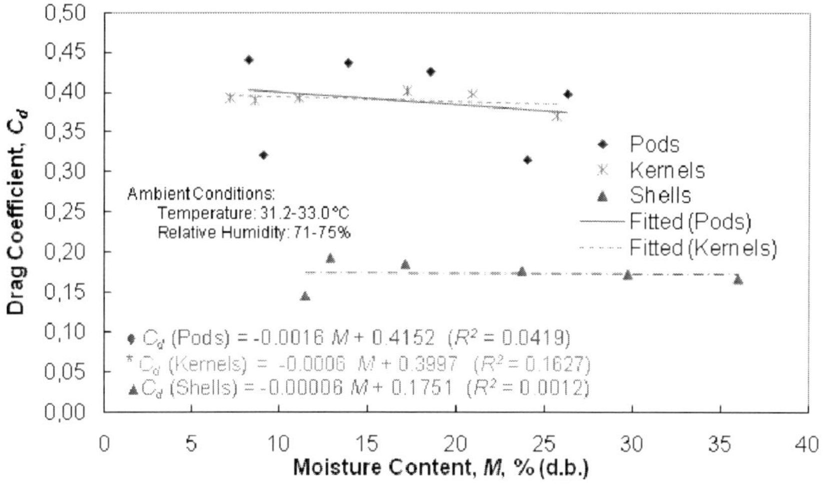


Figure 33. Variation in drag coefficients for pods, kernels, and shells at different moisture contents.

The drag coefficient at various moisture contents was determined using the average terminal velocity data for peanut pods, kernels and shells, and the variation in the average drag coefficient of these materials at different moisture contents is shown in Figure 33. The drag coefficient data was fitted with linear equations. It was found that the drag coefficient for pods, kernels and shells was 0.39, 0.39 and 0.17, respectively. The latter showed that the drag coefficient was equal for peanut pods and kernels. This might be due to a more-or- less equal surface roughness for pods and kernels. In the case of shells, a lower drag coefficient was observed. This might be due to the tremendous swirling of shells in the air stream, possibly resulting in the surface roughness of the shells having much less influence. During the experimentation, a vigorous rotation of shells was observed, compared to pods and kernels, in the transparent Perspex vertical tube.

## AIRFLOW RESISTANCE CHARACTERISTICS

The airflow resistance set-up was designed and developed [12] to study the airflow resistance characteristics of these materials at different moisture contents, airflow rates and depths for the same *cultivar*.

The pressure drops at different heights of airflow resistance set-ups, up to 500 mm, at three pressure taps for each height, were determined and a variance analysis was carried out. It was observed that the average pressure drop of the experimental set-up increased from 0.77 to 1.22 mm of water by increasing the heights from 20 mm to 50 mm. The pressure drop, up to the 20 mm height, represented the set-up loss between the taps arranged on the vertical pipe and the 20 mm taps. Between these two tap locations, one coupling, a conical plenum chamber with three wire screen meshes, a mesh at its top to support the material, and a set-up circular wall of 20 mm height were present. Similarly, the pressure drop, up to a 500 mm height, represented the set-up loss up to 500 mm taps from the taps on the vertical pipe. The difference in the pressure drop between the taps of 20 and 500 mm was 0.4453 mm of water. For an analyzing results, the individual set-up loss was deducted for 20 and 300 mm bed depths for correct results since these bed depths were used for the pressure taps on vertical pipe, whereas, the set-up loss was not considered for other depths because the pressure drop between the taps on the holding bin was negligible. The variance analysis for the data, up to each height at different airflow rates, indicated that the variation in pressure drop was insignificant at 1% and 5% levels of significance. Furthermore, the variance analysis was also conducted for the pooled data for all the depths at different airflow rates and revealed that the variation in pressure drop was insignificant at 1% and 5% levels of significance. Hence, the average pressure drop of 0.9472 mm of water may be used for the set-up, up to a height of 500 mm, with standard deviation of $\pm$0.5521. At each height of the set-up, the variation in pressure drop was insignificant by increasing the airflow rates up to 1.23 $m^3$ $s^{-1}$ $m^{-2}$. Also, the variation in pressure drop, up to a set-up height of 500 mm, was insignificant by increasing the airflow rates up to 1.23 $m^3$ $s^{-1}$ $m^{-2}$ and an average set-up loss of 0.9472 mm of water could be used for calculating the pressure drop in the grain beds.

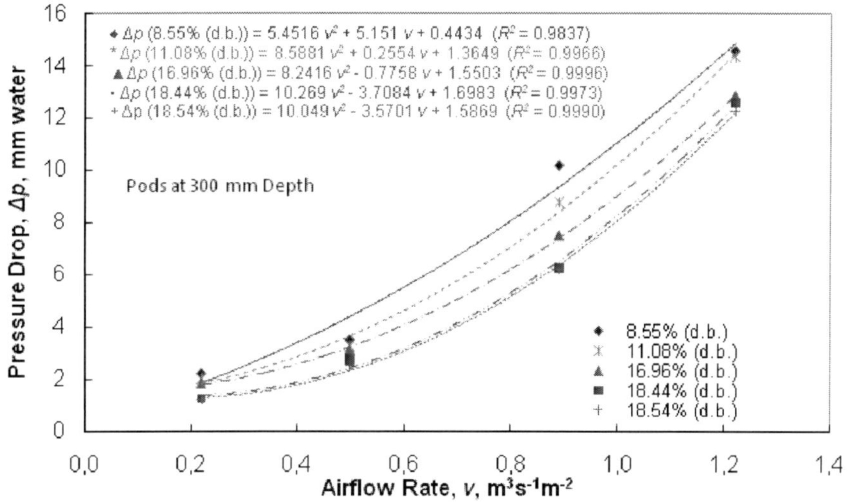


Figure 34. Variation of pressure drop with airflow rate at different m. c.s of pods at a depth of 300 mm.

The experiments were conducted on different depths of peanut pods, kernels and shells (20, 80, 100, 180, 200, 280 and 300 mm) at various moisture contents. It was observed that the pressure drop increased more rapidly with an increase in airflow rate from 0.22 to 1.22 $m^3$ $s^{-1}$ $m^{-2}$ for all the bed depths of pods at each m. c. (Figure 34). At each bed depth, it was found that the pressure drop decreased with an increase in moisture content from 8.55 to 18.54 (d.b.). These results were in accordance with previous research [17, 34] for oilseed rape and lentil seeds. The reduction in the pressure drop with an increase in moisture content was due to a decrease in bulk density and an increase in porosity. Also, it could be observed that the pressure drop had increased by increasing the bed depth of pods at each moisture content and airflow rate. This increase was due to an increase in surface resistance of the material proportionately with bed depths. Further, it was observed that the pressure drop increased by 689 and 601% for pod depths of 20 and 300mm, respectively, by increasing the airflow rate from 0.22 to 1.22 $m^3$ $s^{-1}$ $m^{-2}$ at the moisture content of 16.96% (d.b.). Also, it was observed that the pressure drop decreased by 44 and 39% for pod depths of 20 and 300 mm, respectively, by increasing the moisture content from 8.55 to 18.54% (d.b.) at an airflow rate of 0.89 $m^3\ s^{-1}\ m^{-2}$.

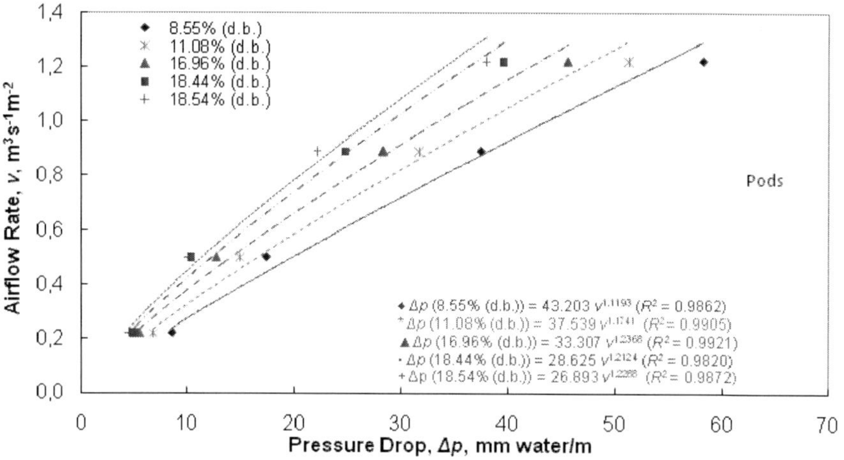


Figure 35. Effect of airflow rate on pressure drop per m depth of pods at different moisture contents.

The pressure drop was calculated for a depth of 1 m for pods by using the data at each depth, and the values were averaged and noted as pressure drop per meter depth. The effect of airflow rate on pressure drop per meter depth for peanut pods as a function of moisture content is shown in Figure 35. It was observed that the pressure drop per m depth increased with increase in airflow rate from 0.22 to 1.22 $m^3\ s^{-1}\ m^{-2}$ at each moisture content. Furthermore, it was found that it decreased with increases in moisture content at each flow rate. This followed the general reported trend [17, 34]. It was observed that the pressure drop per m depth decreased from 37.46 to 22.16 mm of water (30%) when moisture content increased from 8.55 to 18.54% (d.b.) at an airflow rate of 0.89 $m^3\ s^{-1}\ m^{-2}$. Furthermore, the latter increased from 5.44 to 45.59 mm of by water (738%) by increasing the airflow rate from 0.22 to 1.22 $m^3\ s^{-1}\ m^{-2}$ at a pod's moisture content of 16.96% (d.b.). In order to present the pressure drop data in a form suitable for design purposes, its relationship with bulk density, particle density, porosity, moisture content and airflow rate should be determined. In this study, the relationships of physical properties with moisture content were established. The following multiple regression equation was obtained to represent the relationship between pressure drop, $\Delta p$, moisture content, $M$ , and airflow rate, $v$:

$$\Delta p = 0.43002 - 0.36373\ M + 39.59181\ v\ (R^2 = 0.903) \tag{13}$$

For kernels as well, it was observed that the pressure drop increased rapidly with an increase in airflow rate from 0.22 to 1.22 $m^3\ s^{-1}\ m^{-2}$ for all bed depths of kernels at each moisture contents (Figure 36). At each bed depth, it was found that the pressure drop decreased with an increase in moisture content from 9.24 to 18.04% (d.b.) at any airflow rate. It could also be observed that the pressure drop increased by increasing the bed depth of kernels at each moisture content and airflow rate.

Furthermore, it was observed that the pressure drop increased by 1,602 and 1,185% for pod depths of 20 and 300 mm, respectively, by increasing the airflow rate from 0.22 to 1.22 $m^3\ s^{-1}\ m^{-2}$ at a moisture content of 14.69% (d.b.).

In addition, it was observed that the pressure drop decreased by 60 and 25% for bed depths of 20 and 300 mm, respectively, by increasing the moisture content from 9.24 to 18.04% (d.b.) at an airflow rate of 0.50 $m^3\ s^{-1}\ m^{-2}$.

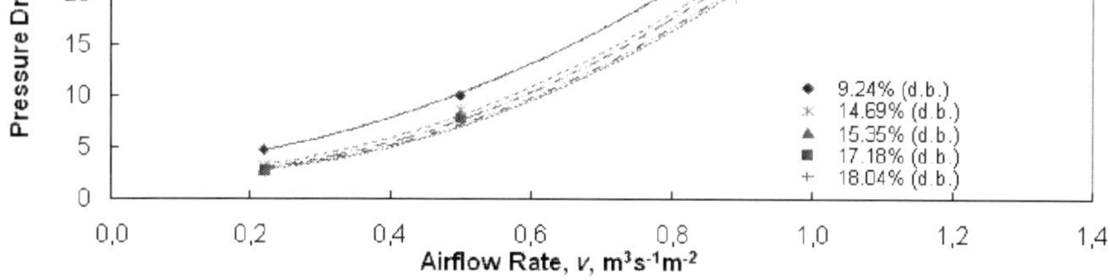


Figure 36. Variation of pressure drops with airflow rates at different m. c.s of kernels at 200 mm depth.

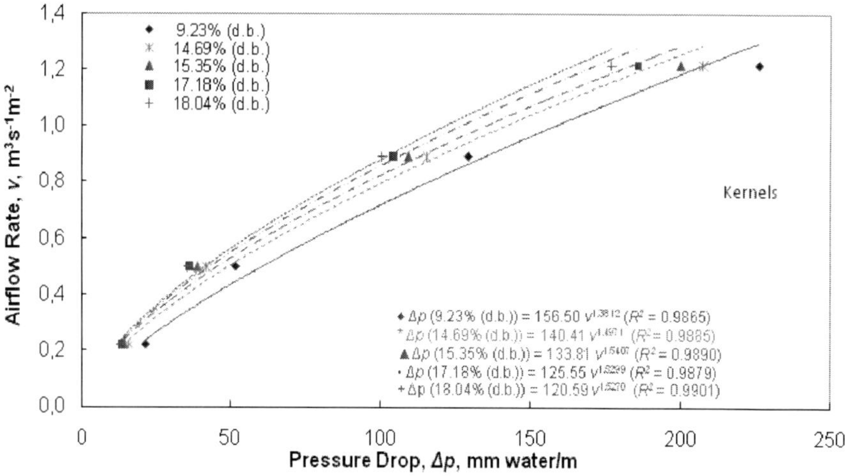


Figure 37. Effect of airflow rate on pressure drop per m depth of kernels at different moisture contents.

The effect associated with the airflow rate of a pressure drop per meter depth of peanut kernels as a function of moisture content is shown in Figure 37. It was observed that the pressure drop per m depth increased with an increase in the airflow rate from 0.22 $m^3\ s^{-1}\ m^{-2}$ at each moisture content as in the case of the pods.

In addition, it was found that the pressure drop per m depth decreased with an increase in moisture content at each flow rate as stated by other researchers. It was observed that the pressure drop per m depth decreased from 51.89 to 35.47 mm of water (32%) when the moisture content increased from 9.23 to 18.04% (d.b.) at an airflow rate of 0.50 $m^3\ s^{-1}\ m^{-2}$. Furthermore, it was found that the pressure drop per m depth increased from 15.91 to 207.18 mm of water (1,202%) by increasing the airflow rate from 0.22 to 1.22 $m^3\ s^{-1}\ m^{-2}$ at a kernel's moisture content of 14.69% (d.b.). The pressure drop was found to be more in the case of kernels compared to pods in all the cases.

This increase might be due to the higher bulk density and lower porosity of kernels compared to those of pods. The following multiple regression equation was obtained to represent the relationship between pressure drop, $\Delta p$, moisture content, $M$, and airflow rate, $v$:

$$\Delta p = 4.62791 - 2.89720\,M + 184.25663\,v\ (R^2 = 0.960). \quad (14)$$

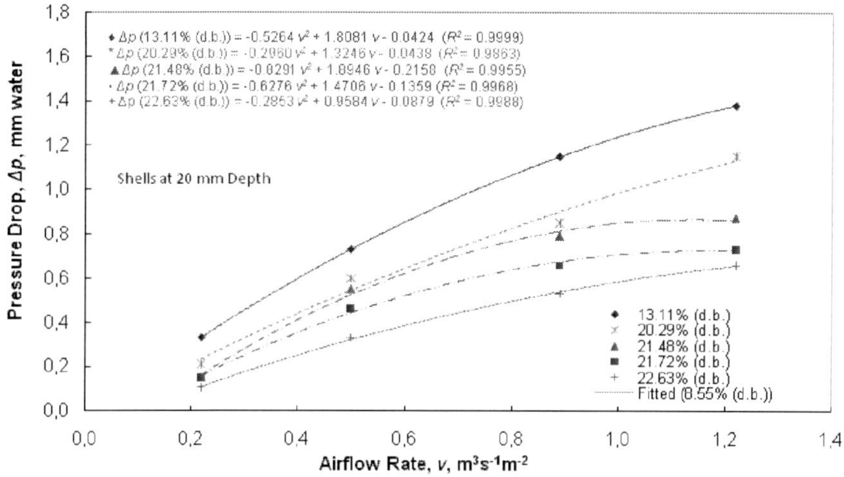


Figure 38. Variation of pressure drops with airflow rates at different m. c. of shells at 20 mm depth.

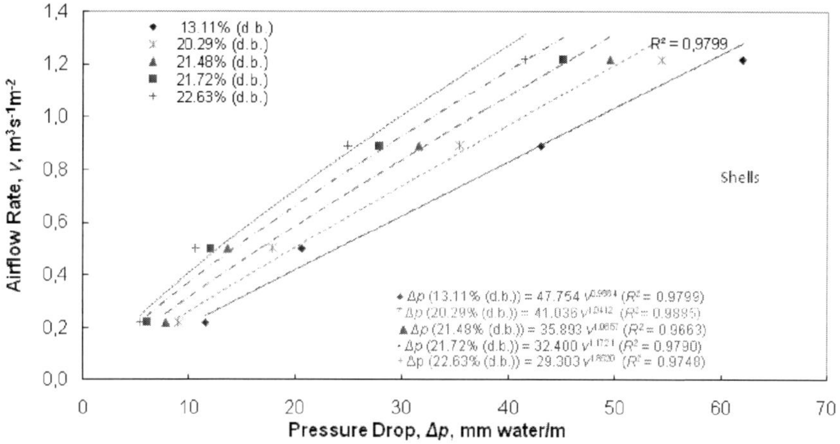


Figure 39. Effect of airflow rates on the pressure drop per m depth for shells at different moisture contents.

In the case of shells, it was observed that the pressure drop increased rapidly with an increase in the airflow rate from 0.22 to 1.22 $m^3$ $s^{-1}$ $m^{-2}$ for all the bed depths of shells at each m.c. (Figure 38). At each bed depth, it was found that the pressure drop decreased with an increase in moisture content from 13.11 to 22.63% (d.b.) at any airflow rate. Also, it could be observed that the pressure drop increased by increasing the bed depth of shells at each moisture content and airflow rate. Furthermore, it was observed that the pressure drop increased by 448 and 614% for pod depths of 20 and 300 mm, respectively, by increasing the airflow rate from 0.22 to 1.22 $m^3$ $s^{-1}$ $m^{-2}$ at a moisture content of 20.29% (d.b.). Also, it was observed that the pressure drop decreased by 52 and 17% for bed depths of 20 and 300 mm, respectively, by increasing the moisture content from 13.11 to 22.63% (d.b.) at an airflow rate of 1.22 $m^3$ $s^{-1}$ $m^{-2}$.

The effect of the airflow rate on the pressure drop per meter depth of peanut shells as a function of moisture content is shown in Figure 39. It was observed that the pressure drop per m depth increased with an increase in the airflow rate from 0.22 to 1.22 $m^3$ $s^{-1}$ $m^{-2}$ at each moisture content, as in the case of pods and kernels. Also, it was found that it decreased with an increase in moisture content at each airflow rate. It was observed that the pressure drop per m depth decreased from 62.00 to 41.48 mm of water (33%) when the moisture content increased from 13.11 to 22.63% (d.b.) at an airflow rate of 1.22 $m^3$ $s^{-1}$ $m^{-2}$. Moreover, it was found that the pressure drop per m depth increased from 9.02 to 54.35 mm of water (502%) by increasing the airflow rate from 0.22 to 1.22 $m^3$ $s^{-1}$ $m^{-2}$ at a kernel's moisture content of 20.29% (d.b.). It was observed that the pressure drop was higher in shell beds than pods and lower than kernels even though the bulk density was less and porosity was more. This trend might be due to the concave portion of the shells offering a higher resistance for airflow. In actual practice, the pressure drop of shells increased marginally compared to pods, even though a higher pressure drop was expected due to the local swirling action of air in the concave portion. This resultant marginal increase in the pressure drop might be due to an expansion of the shells' bed at each airflow rate, to some extent, resulting in a further porosity increase. The following multiple regression equation was obtained to represent the relationship between pressure drop, $\Delta p$, moisture content, $M$, and airflow rate, $v$:

$$\Delta p = 20.80118 - 1.24733\,M + 43.23811\,v \quad (R^2 = 0.961) \qquad (15)$$

## THERMODYNAMIC CHARACTERISTICS

Experiments were conducted [35] on the same *cultivar* to determine variations of the adsorption equilibrium moisture content of peanut pods, kernels and shells by using a static gravimetric method at different equilibrium relative humidities (ERHs) of the environment in glass desiccators and different environment temperatures. The initial moisture contents of peanut pods, kernels and shells were found to be 1.21, 0.98 and 2.30% (d.b.), respectively. These moisture contents were almost similar for all the experiments. This indicated the adsorption of moisture content during equilibrium moisture content (EMC) determination. The data of equilibrium moisture content was analyzed and isotherms were drawn and the data was fitted into ten non-linear isotherm models. The different models used are given in Table 1. These models have three, four or five parameters.

**Table 1. Isotherm models for fitting experimental data**

| Name of the model | Equation |
|---|---|
| Day and Nelson (1965) [36] | $RH = 1 - e^{-a\, t^{b} M^{c\, t^{d}}}$ |
| Modified Henderson (Thomson *et al.*, 1968)[37] | $RH = 1 - e^{-a\,(t+b) M^{c}}$ |
| Modified Halsey (Iglesias and Chirife, 1976) [38] | $RH = e^{\frac{-e^{(a+bt)}}{M^{c}}}$ |
| Modified Chung and Pfost (Pfost *et al.*, 1976) [39] | $RH = e^{\frac{-a\, e^{-b\,M}}{t+c}}$ |
| Modified Oswin (Chen, 1988) [40] | $RH = \frac{1}{\left(\frac{a+b\,t}{M}\right)^{c} + 1}$ |
| Kaleemullah (Kaleemullah, 2004) [41] | $RH = a - b\, e^{-c\, t\, M^{d}}$ |
| Modified Chen (Bitra-I, Proposed) | $RH = \frac{1}{\left(\frac{a+b\,t}{M}\right)^{c} + d}$ |
| Modified Kaleemullah (Bitra-II, Proposed) | $RH = a - b\, e^{-c\, t^{d} M}$ |
| Modified Day and Nelson (Bitra-III, Proposed) | $RH = a - e^{-b\, t^{c} M^{d\, t^{e}}}$ |
| Modified Iglesias and Chirife (Bitra-IV, Proposed) | $RH = e^{\frac{-a\, e^{(b+c\,t)}}{M^{d}}}$ |

The variation in equilibrium moisture content of peanut pods, kernels and shells with equilibrium relative humidities at different temperatures of 35, 50, 65, 80 and 95 °C is shown in Figs. 40, 41 and 42, respectively. The equilibrium moisture content increased with an increase in the ERH for all the temperatures. All the adsorption isotherm curves followed the similar sigmoid shape. A similar trend was observed for green grams [42], sunflower seeds, hulls and kernels [43], ground turmerics [44], osmotic concentrated pineapples [45], Arabian dry cereal foods [46], chilli powder [47], soya beans [48], red chilies [41], mushrooms [49], etc. This might be due to the fact that the vapor pressure from the water present in foods increased with that of the surroundings.

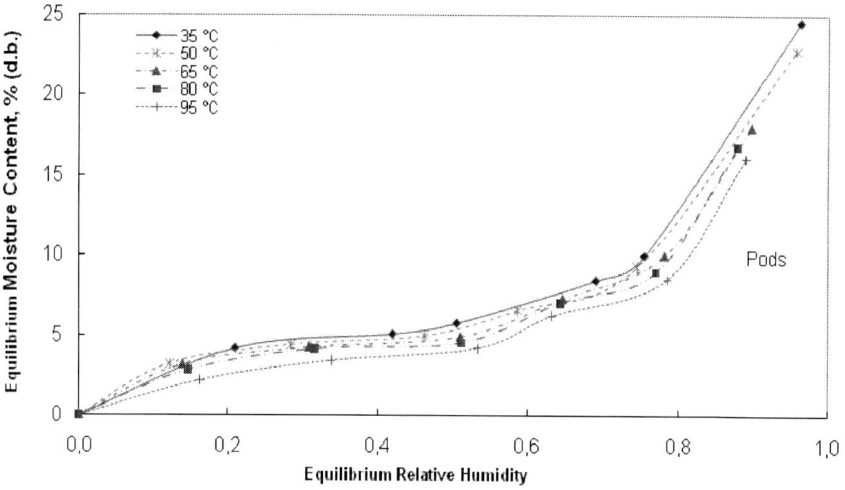


Figure 40. Variation in adsorption EMCs of peanut pods with ERHs at different moisture contents.

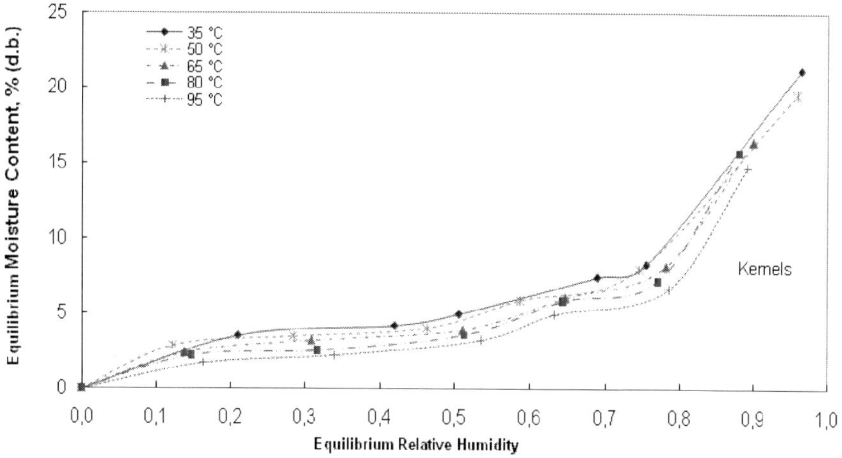


Figure 41. Variation in adsorption EMCs of peanut kernels with ERHs at different moisture contents.

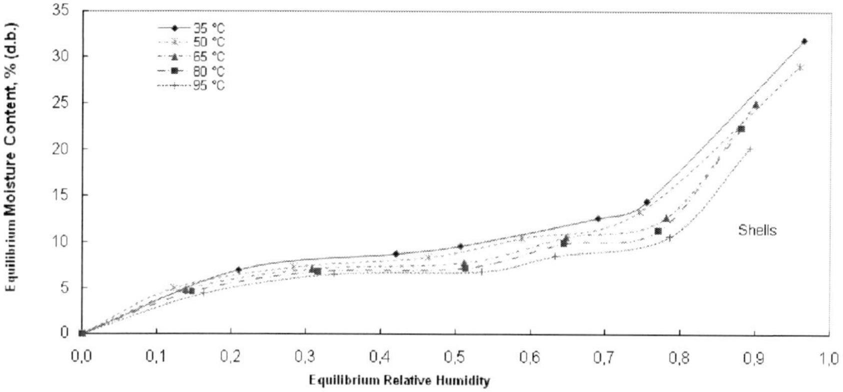


Figure 42. Variation in adsorption EMCs of peanut shells with ERHs at different moisture contents.

The equilibrium moisture content of peanut pods increased from 4.156 to 24.525% (d.b.) with an ERH increase from 0.210 to 0.964 at constant environmental temperature of 35ºC, whereas it increased from 3.22 to 22.75% m.c. (d.b.) with an ERH increase from 0.122 to 0.958 at 50 ºC, from 3.17 to 18.01% m.c. (d.b.) with an ERH increase from 0.139 to 0.898 at 65 ºC, from 2.82 to 16.77% m.c. (d.b.) with an ERH increase from 0.147 to 0.879 at 80 ºC, and from 2.20 to 16.04% m.c. (d.b.) with an ERH increase from 0.163 to 0.891 at 95 ºC. The equilibrium moisture content of peanut kernels increased from 3.52 to 21.23% (d.b.) with an ERH increase from 0.210 to 0.964 at a constant environmental temperature of 35 ºC, whereas it increased from 2.88 to 19.59% m.c.(d.b.) with an ERH increase from 0.122 to 0.958 at 50 ºC, from 2.41 to 16.48% m.c. (d.b.) with an ERH increase from 0.139 to 0.898 at 65 ºC, from 2.24 to 15.17% m.c. (d.b.) with an ERH increase from 0.147 to 0.879 at 80 ºC, and from 1.72 to 14.80% m.c. (d.b.) with an ERH increase from 0.16 to 0.891 at 95 ºC. The equilibrium moisture content of peanut shells increased from 6.96 to 31.95% (d.b.) with an ERH increase from 0.210 to 0.964 at a constant environmental temperature of 35ºC, whereas it increased from 4.96 to 29.08% m.c. (d.b.) with an ERH increase from 0.122 to 0.958 at 50 ºC, from 4.82 to 25.10% m.c. (d.b.) with an ERH increase from 0.139 to 0.898 at 65 ºC, from 4.61 to 22.40% m.c. (d.b.) with an ERH increase from 0.147 to 0.879 at 80 ºC, and from 4.41 to 20.22% m.c. (d.b.) with an ERH increase from 0.163 to 0.891 at 95 ºC. These results indicated that the EMC was higher for shells and was lower for kernelsthan that of the peanut pods at all temperatures and equilibrium relative humidity values. A similar trend was obtained for sunflower seeds, hulls and kernels [43].

The EMC values of peanut pods, kernels and shells decreased with an increase in temperature at all the levels of ERH (Figs. 40, 41, and 42). The reason might be that the vapor pressure of the moisture within the peanut pods, kernels and shells increased and hastened the transfer of moisture from the product to the surrounding air. The kinetic energy associated with water molecules present in the products increased with increasing temperature. This in turn resulted in decreasing the attractive forces, consequently, allowing water molecules to escape. This might have led to a decrease in the degree of water adsorption withan increasing temperature at a given ERH. An upward shift in temperature from 35 to 95 °C led to a shift of isotherms of peanut pods, kernels and shells towards a lower value for the EMC at each ERH, indicating that at any constant equilibrium moisture content, the ERH increased with an increase in temperature.

The adsorption equilibrium moisture content of peanut pods, kernels and shells at various equilibrium relative humidity values and surrounding environment temperatures viz, 35, 50, 65, 80 and 95 °C were modeled by using ten isotherm equations as stated in Table 1. The estimated values of the parameters of these models are given in Tables 2, 3 and 4 for peanut pods, kernels and shells, respectively. An examination of tables showed that the parameter estimates obtained for the ERH models were in accordance with the values in the literature for rough rice [50], soya beans [48] and red chilies [41].

The residuals of the predicted ERH for ten different models were analyzed. Among these models, modified Halsey and modified Iglesias and Chirife (Bitra-IV, Proposed) yielded a random scattering of the residuals in a horizontal band centered on zero, indicating that these two models best fitted the EMC isotherms of peanut pods, kernels and shells.

Various researchers have used the error terms and $R^2$ values for judging the suitability of EMC model equations for safflowers [51], fruits and vegetables [52], soya beans [48], red chilies [41] and mushrooms [49]. Furthermore, in this study, error terms and $R^2$ values were determined [35]. The standard error of estimate, $E_s$, mean relative percent deviation, $E_m$, and coefficient of determination, $R^2$, along with residual plot patterns and ranks, were given. It was found that the modified Iglesias and Chirife (Bitra-IV, Proposed) equation yielded a much less mean percent deviation, $E_m$, of 1.387801, 6.888941 and -1.805685 for pods, kernels and shells, respectively. These values were less than 10% as suggested [53]. Hence, this equation was ranked first. Second best was the modified Halsey equation with less than 10%

of $E_m$ values for all these products. The modified Chung and Pfost and modified Oswin equations gave values less than 10% for peanut pods and peanut shells, respectively, but the residual plots were systematic. Hence, these equations were not considered. It was also observed that the standard error of estimate, $E_s$, values were 0.05055547, 0.06322391 and 0.06168271 for peanut pods, kernels and shells, respectively, for the modified Iglesias and Chirife (Bitra-IV, Proposed) equation. These values were approximately equal to the modified Halsey equation; the latter were much less compared to all other equations.

The variation in the latent heat ratio with moisture content for peanut pods, kernels and shells is shown in Figure 43. The latent heat ratio decreased exponentially with an increase in moisture content for all these materials, and these values were more than unity. A similar trend was also observed for soya beans [48]. At each moisture content, the L/L' ratio was higher for peanut shells, followed by pods and kernels. This might be due to changes in thechemical composition and structural make up. The L/L' ratio decreased exponentially from 1.3447 to 1.0189, from 1.2268 to 1.0118 and from 1.266 to 1.0271 for peanut pods, kernels and shells with an increase in moisture content from 4 to 16% (d.b.), 4 to 14% (d.b.) and 8 to 20% (d.b.), respectively.

The Arrhenius plots for peanut pods, kernels and shells were drawn for different moisture contents by having *ln* ERH on the ordinate and $1/T$ on the abscissa. All these curves had typical nature for an Arrhenius plot. The isosteric heat, $q_{st}$, values at each moisture content for all these materials were calculated from the exponential equations at each m.c. The variation in the isosteric heat of adsorption of peanut pods, kernels and shells with moisture content is shown in Figure 44. The isosteric heat decreased with an increase in moisture content for all these materials. The isosteric heat decreased from 14.698 to 0.804 kJ g-mol$^{-1}$, from 9.612 to 0.502 kJ g-mol$^{-1}$ and 11.289 to 1.154 kJ g-mol$^{-1}$ for peanut pods, kernels and shells with an increase in moisture content from 4 to 16% (d.b.), 4 to 14% (d.b.) and 8 to 20% (d.b.), respectively. These values were in accordance with the turmeric powder [44] and mushrooms [49]. The high $q_{st}$ at low moisture contents might be due to a strong moisture-material interaction in the dried product. As the moisture content increased, the available sites for moisture adsorption might have diminished and resulted in lower $q_{st}$ values. It may also be inferred that the energy requirement for adsorption was higher for peanut shells and followed by pods and kernels.

**Table 2. Estimated values of different parameters of various adsorption models for peanut pods**

| Model | Parameter values | | | | |
|---|---|---|---|---|---|
| | a | b | c | d | e |
| Day and Nelson | 0.0003672529 | 1.035667 | 2.923186 | -0.1557156 | - |
| Modified Henderson | 0.0002813249 | 21.74667 | 1.607554 | - | - |
| Modified Halsey | 3.191663 | -0.009191406 | 1.729503 | - | - |
| Modified Chung and Pfost | 27398.0 | 0.1145405 | 15648.0 | - | - |
| Modified Oswin | 8.928039 | -0.0357187 | 2.340005 | - | - |
| Kaleemullah | 1.047498 | 0.9595861 | 0.0001069881 | 1.933155 | - |
| Modified Chen (Bitr-I, Proposed) | 9.805537 | -0.03543294 | 2.128553 | 0.9347529 | - |
| Modified Kaleemullah (Bitra-II, Proposed) | 1.195378 | 1.354456 | 0.01838138 | 0.3645239 | - |
| Modified Day and Nelson (Bitra-III, Proposed) | 1.035414 | 0.001680092 | 0.5980904 | 1.758583 | -0.02512452 |
| Modified Iglesias and Chirife (Bitra-IV, Proposed) | 0.9193867 | 0.3575751 | -0.009946291 | 1.926373 | - |

## Table 3. Estimated values of different parameters of various adsorption models for peanut kernels

| Model | Parameter values | | | | |
|---|---|---|---|---|---|
| | a | b | c | d | e |
| Day and NelsonModified | 0.001897073 | 0.753736 | 2.281656 | -0.1136017 | - |
| Henderson | 0.0004358578 | 22.54772 | 1.535896 | - | - |
| Modified Halsey | 2.897397 | -0.01249189 | 1.603097 | - | - |
| Modified Chung and Pfost | 257.5691 | 0.2031967 | 15.65978 | - | - |
| Modified Oswin | 7.771897 | -0.03309359 | 2.125493 | - | - |
| Kaleemullah | 0.947594 | 0.9643822 | 0.0005905407 | 1.735 | - |
| Modified Chen (Bitra-I, Proposed) | 10.01073 | -0.03563483 | 1.55482 | 0.8014332 | - |
| Modified Kaleemullah (Bitra-II, Proposed) | 1.346545 | 1.435667 | 0.02617825 | 0.3053003 | - |
| Modified Day and Nelson (Bitra-III, Proposed) | 1.091525 | 0.0008395313 | 0.5844206 | 1.963074 | -0.1551484 |
| Modified Iglesias and Chirife (Bitra-IV, Proposed) | 0.925507 | 2.973147 | -0.01240924 | 1.606733 | - |

**Table 4. Estimated values of different parameters of various adsorption models for peanut shells**

| Model | Parameter values | | | | |
|---|---|---|---|---|---|
| | a | b | c | d | e |
| Day and Nelson | 0.001734484 | 0.4313449 | 2.027534 | -0.04263984 | - |
| Modified Henderson | 0.0000957122 | 21.84141 | 1.820249 | - | - |
| Modified Halsey | 3.831629 | -0.004988171 | 1.817537 | - | - |
| Modified Chung and Pfost | 456.9405 | 0.1507063 | 38.93009 | - | - |
| Modified Oswin | 11.42747 | -0.0242201 | 2.453853 | - | - |
| Kaleemullah | 0.9310653 | 0.9101961 | 1.064151 | 1.915138 | - |
| Modified Chen (Bitra-I, Proposed) | 11.57327 | -0.01915862 | 2.314964 | 0.9670351 | - |
| Modified Kaleemullah (Bitra-II, Proposed) | 1.2459 | 1.480851 | 0.01985935 | 0.2664479 | - |
| Modified Day and Nelson (Bitra-III, Proposed) | 1.043598 | 0.0006269759 | 0.5494009 | 1.83056 | -0.004142822 |
| Modified Iglesias and Chirife (Bitra-IV, Proposed) | 0.9158943 | 3.892094 | -0.004910234 | 1.804604 | - |

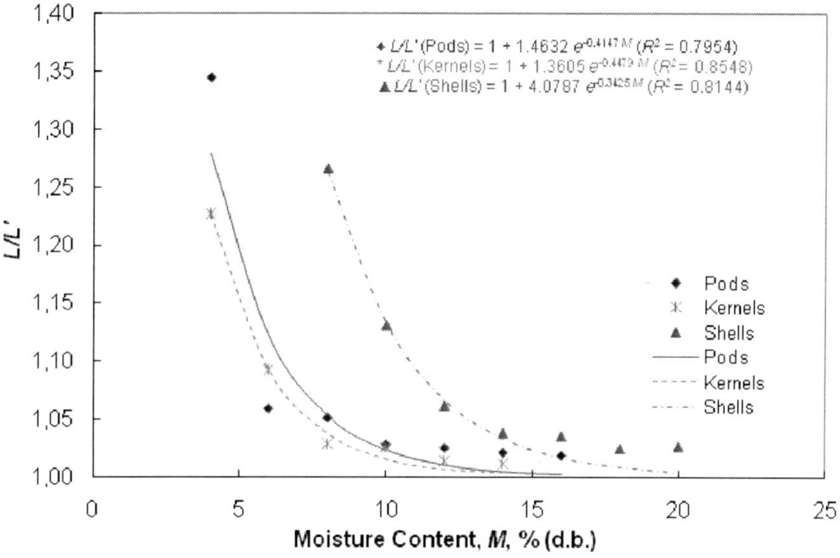


Figure 43. Variation in latent heat ratios of pods, kernels and shells at different moisture contents.

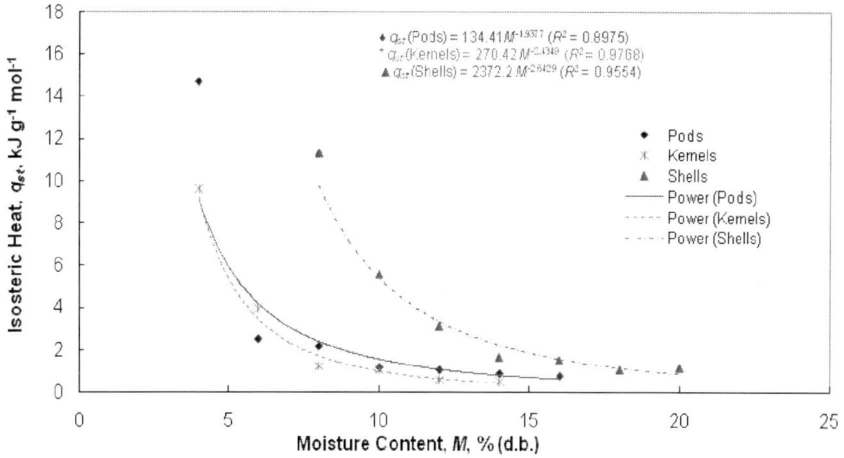


Figure 44. Variation in isosteric heat of adsorption for pods, kernels and shells with moisture content.

## THERMAL PROPERTIES

The specific heat, thermal conductivity, and thermal diffusivity, of peanut pods, kernels, and shells were evaluated as a function of moisture content for the same *cultivar* [54]. The specific heat and thermal conductivity were measured using a purposely built vacuum flask calorimeter and a transient-state heat transfer apparatus, respectively. The specific heat was determined by mixing hot water with a sample maintained at ambient temperature in the calorimeter and incorporating temperature-correction into the calculations. Thermal conductivity was determined in 60-minute experiments after checking for time-correction in the calculations. Thermal diffusivity was calculated by using formula method.

The variation in specific heat, thermal conductivity and thermal diffusivity with moisture content for peanut pods, kernels and shells is shown in Figure 45, 46, and 47, respectively. The specific heat of the samples increased linearly with increasing moisture content (Figure 45). Specific heat increased from 2.1 to 3.3, 1.9 to 2.8, and 2.7 to 4.1 kJ $kg^{-1}$ $°C^{-1}$ for peanut pods, kernels, and shells with an increase in moisture content from 5.2 to 23.7, 5.0 to 30.6, and 3.5 to 28.7% (d.b.), respectively. The valid mid-temperatures averages between ($t_e$+$t'$) and $t_s$, at which the determined specific heat values were applicable. These temperatures were greater or less than the valid specific heat temperature for the calorimeter (50.5 °C). Hence, the specific heats of the samples measured, using the specific heat value of the calorimeter, were quite reasonable. The specific heat increased by 58, 48, and 50% from 2.1, 1.9, and 2.7 kJ $kg^{-1}$ $°C^{-1}$ for peanut pods, kernels, and shells, respectively. Furthermore, the specific heat of shells was the highest, followed by pods and kernels. The increased specific heat of the shells was attributed to the actual chemical composition of the shells. Hence, pods (kernels + shells) stood with a lesser specific heat than shells, whereas kernels (pods – shells) had the lowest specific heat. Earlier researchers also observed an increased specific heat with an increase in moisture content for peanut pods, ground shells, and whole and ground kernels [55]. The increased specific heat with an increase in moisture content might be due to the high specific heat value of water compared to the dry material having occupied its air-filled pores with lower moisture. Hence, these results indicated a reduced specific heat for bone-dry materials (pods: 1.6 kJ $kg^{-1}$ $°C^{-1}$, kernels: 1.7 kJ $kg^{-1}$ $°C^{-1}$; shells: 2.6 kJ $kg^{-1}$ $°C^{-1}$) compared to water (4.183 kJ $kg^{-1}$ $°C^{-1}$ at 50°C). The sample specific heat at any moisture content can be estimated from the linear relationships between specific heat and moisture content. All these relationships showed mean relative percent deviation (*MRPD*) values < 10%, suggesting goodness of fit.

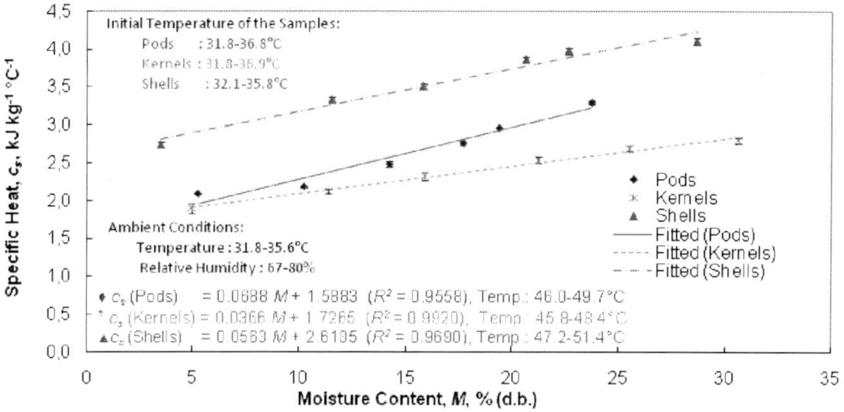


Figure 45. Variation in the specific heat of peanut pods, kernels and shells with moisture content.

The thermal conductivity of samples increased linearly with increasing moisture content (Figure 46). Thermal conductivity increased from 0.12 to 0.16, 0.15 to 0.19, and 0.11 to 0.18 W $m^{-1}$ $°C^{-1}$ for peanut pods, kernels, and shells with an increase in moisture content from 5.2 to 23.7, 5.0 to 30.6, and 3.5 to 28.7% (d.b.), respectively. Increased thermal conductivity with an increase in moisture content for peanut pods, ground shells, and whole and ground kernels was also reported [55]. Thermal conductivity increased by 31%, 24%, and 63% from 0.12, 0.15, and 0.11 W $m^{-1}$ $°C^{-1}$ for peanut pods, kernels, and shells, respectively. The thermal conductivity of shells was the lowest, followed by pods and kernels. The increased thermal conductivity with an increase in moisture content might be due to a higher thermal conductivity of water (0.628 W $m^{-1}$ $°C^{-1}$ at 38 °C) compared to the sample's dry material, associated with air-filled pores. The bulk density of the peanut shells was the lowest, followed by pods and kernels. Furthermore, the porosity was the highest for the peanut shells, followed by pods and kernels [12]. Therefore, porosity decreased as the bulk density increased, which resulted in the highest thermal conductivity for the kernels, followed by the pods and shells. The thermal conductivity of bone-dry peanut pods, kernels, and shells was 0.11, 0.15, and 0.010 W $m^{-1}$ $°C^{-1}$, respectively. The thermal conductivity of samples at any moisture content can be calculated from linear equations (Figure 46).

The *MRPD* was less than 10%, indicating a goodness of fit for these regression equations.

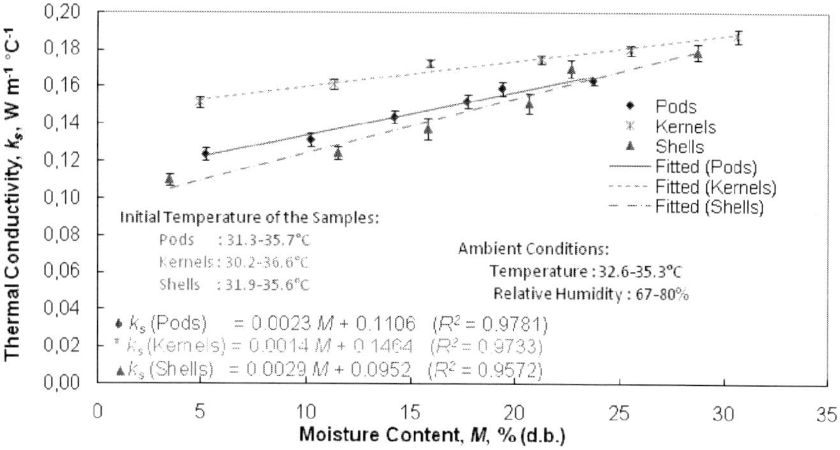


Figure 46. Variation in thermal conductivities for peanut pods, kernels and shells with moisture content.

Thermal diffusivity decreased from $2.8\times10^{-7}$ to $2.3\times10^{-7}$ and $1.1\times10^{-7}$ to $1.0 \times10^{-7}$ m$^2$ s$^{-1}$ for peanut pods and kernels, respectively, and it increased from $5.9\times10^{-7}$ to $6.7\times10^{-7}$ m$^2$ s$^{-1}$ for shells with increasing moisture content from 5.2 to 23.7, 5.0 to 30.6, and 3.5 to 28.7% (d.b.), respectively (Figure 47). Thermal diffusivity decreased by 16% and 13% from $2.8\times10^{-7}$ and $1.1\times10^{-7}$ m$^2$ s$^{-1}$ for peanut pods and kernels, respectively, and it increased by 14% from $5.9\times10^{-7}$ m$^2$ s$^{-1}$ for shells. The thermal diffusivity of shells was the highest, followed by pods and kernels. In the case of shells, the increased thermal diffusivity with an increase in moisture content might be due to the higher thermal conductivity of water (0.628 Wm$^{-1}$°C$^{-1}$ at 38 °C) compared to the dry material of shells associated with air-filled pores. The increased thermal diffusivity of shells might be due to its poor heat conduction. Earlier researchers observed a decreased thermal diffusivity with increased moisture content for rapeseed [56] and maize [57]. Peanut pods and kernels, which are oil bearing materials, like rapeseed, showed decreased thermal diffusivity with an increased moisture content. The thermal diffusivity of bone-dry peanut pods, kernels, and shells was 3.002, 1.151, and 5.440 Wm$^{-1}$°C$^{-1}$, respectively. The thermal diffusivity of pods, kernels, and shells at any moisture content can be accurately determined from linear relationships (Figure 9). The *MRPD* was $< 10\%$, indicating a goodness of fit for these linear equations.

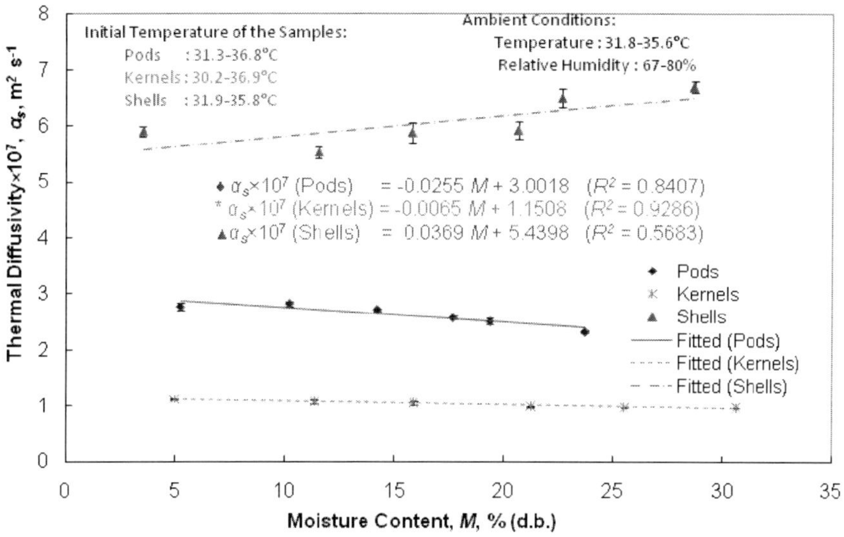


Figure 47. Variation in thermal diffusivities for peanut pods, kernels and shells with moisture content.

## References


[1] B. B. Desai, P. M. Kotecha and D. K. Salunkhe, *Science and Technology of Groundnut: Biology, Production, Processing and Utilization*, Naya Prokash, Calcutta (1999).

[2] S. K. Sharma, R. K. Dubey and C. K. Teckchandani, *Proceedings of the Indian Society of Agricultural Engineers Silver Jubilee Convention*, Vol. 3 - Process, Food and Dairy Engineering. p.II-181 - II-185 (1985).

[3] N. N. Mohsenin, *Physical Properties of Plant and Animal Materials,* 2nd Ed. Gordon and Beach publishers, N.Y (1986).

[4] J. E. Chapman, R. V. Morey, H. A. Cloud and J. L. Nieber, *Transactions of the ASAE* 32(4),1368-1376 (1989).

[5] S. Szcezesniak, *Physical Properties of Food.* Ed. by M. Peleg and E. B. Bagley, AVI Publishers, Connecticut. p. 7-16 (1983).

[6] K. Alagusundaram, D. S. Jayas, W. E. Muir and N. D. G. White, *Transactions of the ASAE* 34(4), 1784-1788 (1991).

[7] S. Deepika and V. N. L. Lavanya, Studies on Dimensional Characterization, Aerodynamic and Frictional Properties of Peanut Pods,

Kernels and Husk, Unpublished B.Tech. (Agricultural Engineering) Project Report, *ANGRAU*, A. P., India (2002).

[8] S. H. Adhaoo, S. T. Ingle, S. U. Dhawale, *Journal of Agricultural Engineering* 13 (1), 19-21 (1976).

[9] K. K. Singh, T. K. Goswami, *Journal of Agricultural Engineering Research* 64, 93-98 (1996).

[10] H. Öğüt, *Journal of Agricultural Engineering Research* 69, 273-277 (1998).

[11] S. N. Jha, *Journal of Agricultural Engineering Research* 72, 145-150 (1999).

[12] G. Sreedhar and V. V. Rao, Some Studies on Physical Properties and Airflow Resistance Characteristics of Peanut Pods, Kernels and Husk, Unpublished B.Tech. (Agricultural Engineering) Project Report, *ANGRAU*, A. P., India (2001).

[13] B. P. N. Singh, P. C. Sah, and A. K. Tikkoo, *Engineering Properties of Food Materials*. All India Co-ordinated ICAR Scheme for Studies on Harvest and Post-Harvest Technology, CIAE, Bhopal (1980).

[14] D. K. Dev, P. N. Satwadhar and U. M. Ingle, *Journal of Agricultural Engineering* 19(2), 37-48 (1981).

[15] C. R. Sukumaran, *Load Determination and Oil Expression Characteristics of Rapeseed (Brasica compelis) Under Uniaction Bulk Compression.* Unpublished Ph.D. Thesis, G.B. Pant University of Agricultural and Technology, Pantnagar, U. P., India (1983).

[16] V. V. Sreenarayanan, V. Subramaniyan, and R. Visvanathan, *Proceedings of ISAE, Silver Jubilee Convention*, Vol. 3, pp.161-169 (1985).

[17] B. G. Patil and G. T. Ward, *Journal of Agricultural Engineering Research,* 41(1), 25-31 (1988).

[18] D. S. Jayas, S. Sokhansanj and N. D. G. White, N.D.G. *Transactions of the ASAE* 32(1), 291-294 (1989).

[19] W. L. McCabe, J. C. Smith and P. Harriott, P. *Unit Operations of Chemical Engineering,* 5th Ed. McGraw-Hill, Inc., N. Y. (1993).

[20] S. Nath and A. K. Inima, A.K. *Proceeding of ISAE Silver Jubilee Convention,* Vol. 3, pp.17-26 (1985).

[21] N. Bharathi and V. V. S. Kumar, Studies on Frictional Properties of Peanut Pods, Peanut Kernels, Peanut Husk, Black Gram and Rough Rice, Unpublished B.Tech. (Agricultural Engineering) Project Report, *ANGRAU*, A. P., India (2003).

[22] J. H. Chung and L. R. Verma, *Transactions of the ASAE* 32(2), 745-750 (1989).
[23] R. P. Kachru and D. R. Rai, Physical constituents and some engineering properties of peanut pod. *Journal of Oilseeds Research* 10(2): 264-270 (1993).
[24] K. Varshney, J. N. Nandasana and R. M. Satasiya, *Gujarat Agricultural University Research Journal* 20 (1), 129-133 (1994).
[25] M. A. Hossain, M. A. Khaleque and M. S. Hassan, *Legume Research* 21 (2): 79-84 (1998).
[26] M. S. Teotia, P. Ramakrishna, S. K. Berry and S. Kaur, *Journal of Agricultural Engineering Research* 9, 153-162 (1988).
[27] Sadhana, *Indian Journal of Agricultural Engineering* 1(1), 17-22 (1991).
[28] D. C. Joshi, S. K. Das and R. K. Mukherjee, *Journal of Agricultural Engineering Research* 54, 219-229 (1993).
[29] K. çarman, *Journals of Agricultural Engineering Research* 63, 87-92 (1996).
[30] S. H. Suthar and S. K. Das, *Journal of Agricultural Engineering Research* 65, 15-22 (1996).
[31] R. K. Gupta and S. K. Das, *Journal of Agricultural Engineering Research* 66, 1-8 (1997).
[32] C. J. M. Tado, P. Klacker, H. D. Kutzbach and D. C. Suministrado, *Agricultural Engineering Journal* 8(2), 91-100 (1999).
[33] P. M. Nimkar and P. K. Chattopadhyay, *Journal of Agricultural Engineering Research* 80(2), 183-189 (2001).
[34] S. Sokhansanj, A. A. Falanciski, F. W. Sosulki, D. S. Jayas and J. Tang, J. *Transactions of the ASAE* 33(4), 1281-1285 (1990).
[35] D. Prasad, B. V. Goud and G. C. Chakravarthy, Some Studies on Moisture Adsorption Isotherms of Peanut Pods, Kernels and Husk, Unpublished B.Tech. (Agricultural Engineering) Project Report, *ANGRAU*, A. P., India (2005).
[36] D. L. Day and G. L. Nelson, *Transactions of the ASAE* 8(2): 293-297 (1965).
[37] T. L. Thomson, R. M. Pearf and G. H. Foster, *Transactions of the ASAE* 24(3), 582-586 (1968).
[38] H. A. Iglesias and J. Chirife, *Journal of Food Technology* 11(2), 109-116 (1976).
[39] H. B. Pfost, S. G. Mourer, D. S. Chung and G. A. Miliken, *ASAE Paper* No. 76-3520, St. Joseph, MI (1976).

[40] C. Chen, *A Study of Equilibrium Relative Humidity for Yellow-Dent Corn Kernels*. Unpublished Ph.D. Thesis, University of Minnesota, St. Paul, USA (1988).
[41] S. Kaleemullah and R. Kailappan, *Journal of Biosystems Engineering* 88(1), 95-104 (2004).
[42] K. N. Patil and P. C. Grover, *Journal of Agricultural Engineering* 23(4), 347-357 (1986).
[43] G. Mazza and D. S. Jayas, *Transactions of the ASAE* 34(2), 534-538 (1991).
[44] V. S. Pawar, D. K. Dev, V. D. Pawar, A. B. Rodge, V. D. Surve and D. R. More, *Journal of Food Science and Technology* 29(3), 170-173 (1992).
[45] S. M. Pokharkar and S. Prasad, *Journal of Food Science and Technology* 35(6), 518-520 (1998).
[46] K. R. Kumar, *Journal of Food Science and Technology* 37(3), 330-333 (2000).
[47] V. Munde, *Journal of Maharashtra Agricultural University* 25(1): 55-59 (2000).
[48] N. A. Aviara, O. O. Ajibola and S. A. Oni, 2004. *Journal of Biosystems Engineering* 87(2), 179-190 (2004).
[49] U. S. Shivhare, S. Arora, J. Ahmed and G. S. V. Raghavan, *Lebensmittel Wissenshaft und Technologie* 37, 133-137 (2004).
[50] C. Chen, C. 2001. *Transactions of the ASAE* 44(5), 1241-1245 (2001).
[51] D. S. Jayas and G. Mazza, *Transactions of the ASAE* 34(5), 2099-2103 (1991).
[52] Y. Soysal and S. Öztekin, *Agricultural Mechanization in Asia, Africa and Latin America* 32(2), 19-22 (2001).
[53] C. P. McLaughlin and T. R. A. Magee, *Journal of Food Engineering* 35, 267-280 (1998).
[54] V. S. P. Bitra, S. Banu, P. Ramakrishna, G. Narender, A. R. Womac, *Biosystems Engineering* 106, 503-512 (2010).
[55] D. A. Suter, K. K. Agrawal and V. I. Clary, *Transactions of the ASAE,* 18(2), 370-375 (1975).
[56] W. K. Bilanski and D. R. Fisher, *Transactions of the ASAE* 19(4), 788-791 (1976).
[57] R. C. Verma and S. Prasad, *Journal of Food Science and Technology* 37(5), 500-505 (2000).

In: Peanuts: Bioactivities and Allergies ISBN: 978-1-62417-387-5
Editors: I.L. Expósito, A.B. Blázquez 

*Chapter 3*

# THE PEANUT USED AS A SOURCE OF PHYTOCHEMICALS

*A. O. Cherif**
Laboratoire de Biochimie, des Lipides et des Protéines,
Département de Biologie, Faculté des Sciences de Tunis, Tunis, Tunisia

## ABSTRACT

Plant derived phytochemicals have been the focus of recent research due to their health promoting effects. Phytochemicals are bioactive non-nutrient chemical compounds found in plant foods, such as fruits, vegetables and grains which may be potent effectors of biological processes and have the capacity to influence disease risk via several complementary and overlapping mechanisms.

*Arachis hypogaea*, known as the peanut, is native to South America. The different species are cultivated in Asia, Africa and America, mainly for high-quality vegetable oil production but also as a feedstock and as *in natura* or processed food for human consumption. Beyond its nutritional characteristics and commercial value, peanuts have also been identified as a source of phytochemicals such as trans-resveratrol, phytosterols, policosanol, isoflavones and squalene. These biologically active compounds have been reported to possess biological effects associated with a reduced disease-risk.

* E-mail address: aicha.echerif @gmail.com.

This chapter aims to give an overview of the different uses of the peanut as bioactive food, focusing on the constituents of peanut and its biological activities.

## INTRODUCTION

The traditional use of the peanut for medicinal purposes has been reported since ancient times. In 2003, a qualified health claim was approved, stating that eating 42g of nuts per day could reduce the risk of heart disease [1]. Peanut skins are used to treat chronical hemorrhage and bronchitis in China [2]. Furthermore, peanut extracts have been used in the management of diabetes patients in northern Nigeria. In addition, *Arachis* is used to lower cholesterol, aid weight loss, and prevent cardiovascular diseases and cancer [3]. Also, the leaves of A. *hypogaea* have an astringent action and several biological properties. They are used therapeutically against abdominal pain, bronchitis, constipation, and flatulence. The peanut also has therapeutical effects as a solvent for bleeding in hemophiliacs [4].

Peanuts have shown to have a favorable nutrient profile, presenting several highly valued dietary constituents, including dietary fibers, proteins, micronutrients, and phytochemicals such as phytosterols, phenolics, stilbenes, and arginine [5,6]. Phytochemicals are the bioactive, non-nutrient, naturally occurring plant compounds found in fruits, vegetables, and whole grains [7]. These biologically active compounds have been reported to elicit several biological effects, including cardioprotective, anti-inflammatory, anticancer and others. Although the biological activity of pure compounds has been proved, the intake of peanuts or their extract may sometimes be more favorable than the ingestion of pure phytochemicals. Actually, epidemiological and intervention studies have shown that the frequent consumption of peanuts promotes cardiovascular health by lowering serum low density lipoprotein (LDL)-cholesterol levels and reduces the risk of type II diabetes [8]. They have been also shown to promote weight management when consumed as part of a low-fat diet as a result of its satiating effect [9]. Further, some phytochemicals undergo bacterial modification that produce metabolites that are more biologically active than the parent compounds [10].

Organoleptic properties of peanuts vary among cultivars, according to the oil content, and roasting conditions [11]. Also, agronomic and postharvest factors have been shown to influence the chemical composition and sensory properties of peanuts [11].

In this chapter, a discussion regarding the various uses of the peanut as a bioactive food, focusing on its active constituents and biological activities will be carried out. The following list of components will be considered:

- Bioactive fatty acids.
- Minor biologically active components (tocopherols, stilbenes, phenolic acids, isoflavones, phytosterols, policosanol, and squalene).

## BIOACTIVE FATTY ACIDS

The peanut kernel contains ~ 52% oil (by weight). 98% to 99% of this is triglycerides and 1% to 2% are minor components. Among the triglycerides, the main fatty acids are represented by monounsaturated (MUFA) (oleic 36-67%) with a slight amount of saturates (palmitic, stearic 11%) and an adequate presence of polyunsaturated fatty acids (PUFA) ω6 (linoleic 15-23%).

Peanut, peanut oil, and peanut butter from six varieties of *A. hypogaea* from Nigeria and two from Turkey also have significant levels of capric, lauric, myristic, palmitoleic, and linolenic acids (ω3) [12]. The substitution of diets rich in saturated fats, by diets with a high content in MUFA and PUFA (like peanut oil), can reduce an individual's risk of cardiovascular disease by leading to lower LDL-cholesterol, lower serum triglycerides, by maintaining high-density lipoprotein cholesterol levels in human blood plasma [13].

Moreover, a higher intake of PUFA from peanuts, peanut butter, and peanut oil may improve insulin sensitivity and reduce the risk of developing type-2 diabetes in women [13]. Those effects may be more pronounced in the case of high-oleic peanuts (80% oleic and 5% linoleic) due to the greater presence of MUFA. The high oleic level provides peanuts with potentially greater health benefits and serves to prolong shelf life characteristics [14]. Additionally, peanuts were found to lack naturally occurring trans-fatty acids, which have been adversely associated with various diseases [15].

## TOCOPHEROLS

The major groups of phytochemicals that may contribute to the total antioxidant capacity (TAC) of plant foods include polyphenols, carotenoids and the traditional anti-oxidant vitamins such as vitamin C and vitamin E (tocopherols).

Tocopherols consist of a polar chromanol group and an apolar isoprenoid side chain. The isoprenoid tail of the tocopherols is fully saturated and generically known as vitamin E (Figure 1).

Functionally, vitamins E are potent antioxidants that protect plant membranes against photoinhibition and other photooxidative stresses [16].The major physiological role of tocopherols in the plant is to prevent lipid peroxidation during seed dormancy, germination and early seedling development [16].

Peanuts are a good source of tocopherols. The content of tocopherol in peanuts changes according to the variety and the production location. Peanut oil mainly contains α-tocopherol (50-373 ppm) and γ-tocopherol (90-390ppm) [17].Tocopherols are widely used for food, feed, pharmaceuticals, cosmetics and resins. In food, they are antioxidants in frying oil, margarine and fried snacks. Vitamin E prevents oxidation of body lipids including PUFA, in atherosclerosis and lipid components of cells and organelle membranes. Also, it has been associated with the reduction of heart disease , the delay in Alzheimer's disease, and cancer prevention.

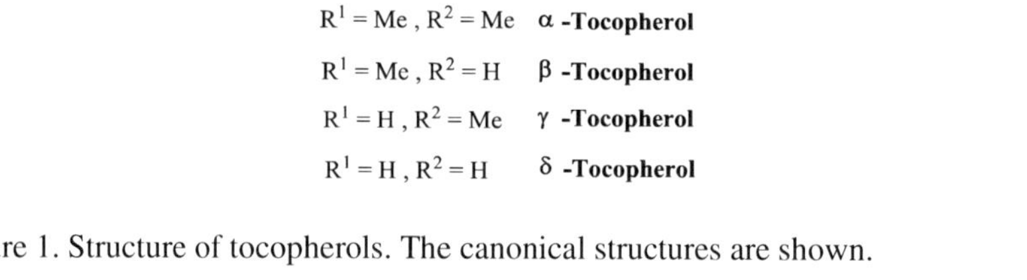


$R^1$ = Me , $R^2$ = Me **α -Tocopherol**

$R^1$ = Me , $R^2$ = H **β -Tocopherol**

$R^1$ = H , $R^2$ = Me **γ -Tocopherol**

$R^1$ = H , $R^2$ = H **δ -Tocopherol**

Figure 1. Structure of tocopherols. The canonical structures are shown.

# POLYPHENOLS

The polyphenols are some of the most studied compounds and can be further divided into stilbenes, flavonoids (including flavonols, flavones, catechins, flavanones, anthocyanidins, and isoflavones), phenolic acids, coumarins and tannins. They will be extensively revised in chapter 4.

## STILBENES

Since resveratrol is considered to be involved in the health benefits associated with a moderate consumption of red wine, plant stilbenes have received notable interest [18]. Stilbenes are characterized by a 1,2-diphenylethylene backbone, usually derived from the basic unit *trans*-resveratrol (3,5,4'-trihydroxy-*trans*-stilbenes) (Figure 2), although other structures are found in particular plant families [19]. Stilbenes are synthesized by a wide range of plant species; however, they are often found in plants that are not routinely consumed as food or in edible seeds. Peanuts (1.3-3.7 $\mu$g of resveratrol/g) and peanut butter are considered a major dietary source of stilbenes, along with grapes and their derivatives [20].

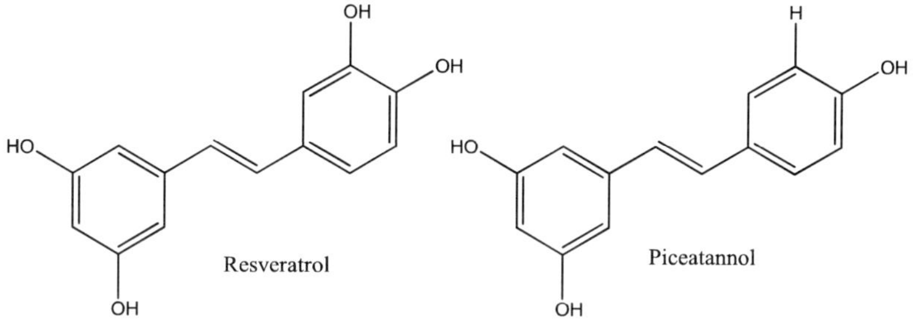


Figure 2. *Trans*-resveratrol and *trans*-piceatannol Structures.

Stilbene synthesis is produced by injuries, fungal contamination, insect damage, and other attacks [21, 22]. Although it also can be found in uninfected and uninjured plants, albeit in minor amounts.

The stilbenes that have been reported in several varieties of A.*hypogaea* in different organs of the plant (hairy roots, leaves, roots, seeds) and in spores (*Aspergillus flavus and Aspergillus parasiticus),* seem to be derived from *trans*-resveratrol, piceatannol, piceid, arachidin-1, arachidin-2, arachidin-3, and *trans*-SB-1 (Figure 3). Nevertheless, after an experiment using *Aspergillus caelatus* spores to induce phytoalexine production in peanuts, other stilbenes such as chiricanine A, arahypin-1, arahypin-2, arahypin-3, arahypin-4 and arahypin-5 were reported [12].

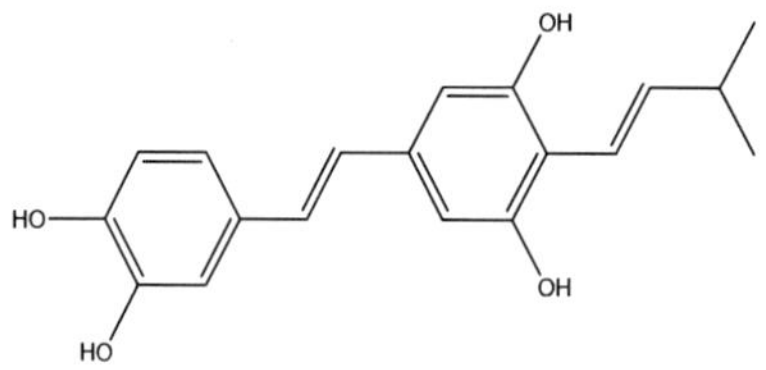

arachidin-1

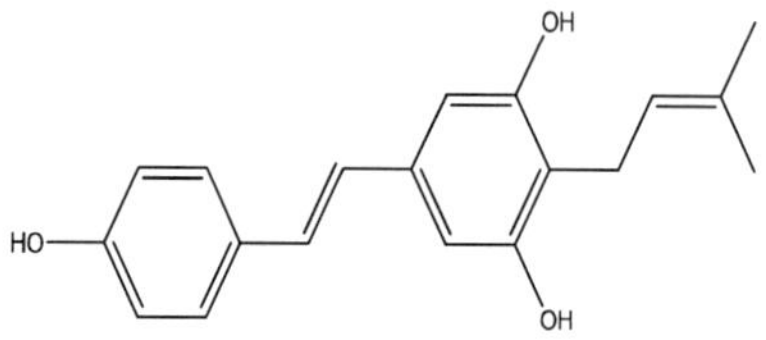

arachidin-2

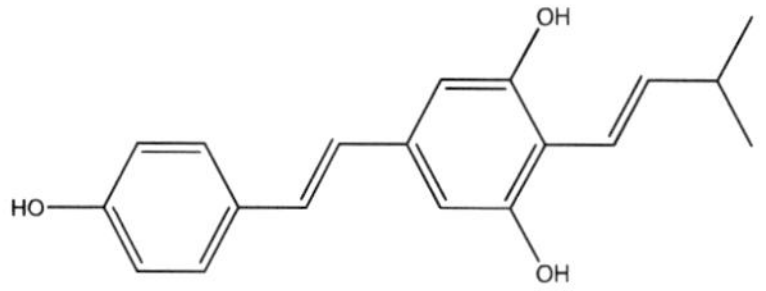

arachidin-3

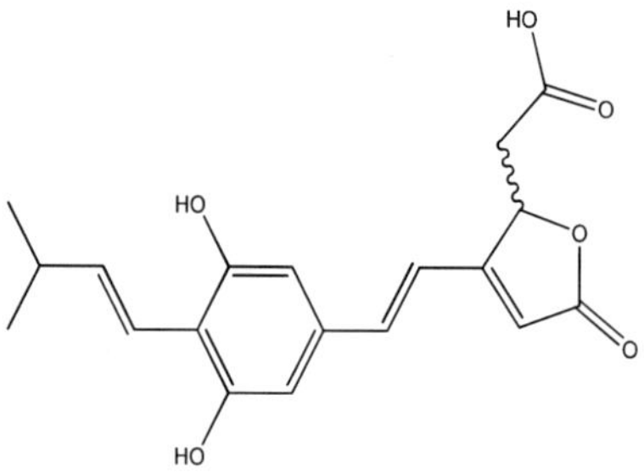

*trans*-SB-1

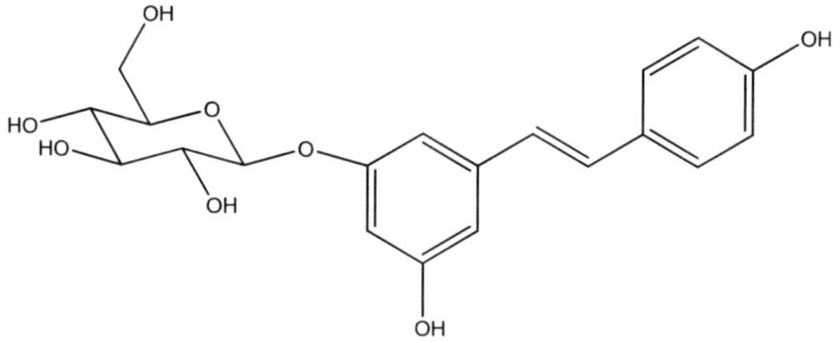

piceid

Figure 3. Arachidin-1, arachidin-2, arachidin-3, *trans*-SB-1 and piceid structures.

These phytochemical components have a free radical scavenging activity that may be related to the prevention of the development of chronical diseases [23]. Piceatannol, in particular, is described as an efficient inducer of apoptosis [24]. Others reports, suggested that both resveratrol and piceatannol are able to induce apoptosis in many cancer- cell lines, but to various extents [24]. In fact, Lin and al [24], reported that piceatannol is more effective in cancer prevention and could be an anticancer compound, resveratrol being a chemoprotective one. Nowadays, resveratrol and piceatannol are becoming promising natural compounds in cancer treatments because they block the carcinogenesis stages [25].

## FLAVONOIDS

The term "flavonoid" is generally used to describe a broad variety of natural products that include a $C_6$-$C_3$-$C_6$ carbon framework, or more specifically a phenylbenzopyran functionality.

Depending on the position of the linkage of the aromatic ring to the benzopyrano moiety, this group of natural products may be divided into three classes: Flavonoids (2-phenylbenzopyrane derivatives, i.e. flavone), Isoflavonoids (3- phenylbenzopyrane derivatives, i.e. isoflavone) and neoflavonoids (4-phenylbenzopyrane derivatives, i.e. neoflavone) (Figure 4).

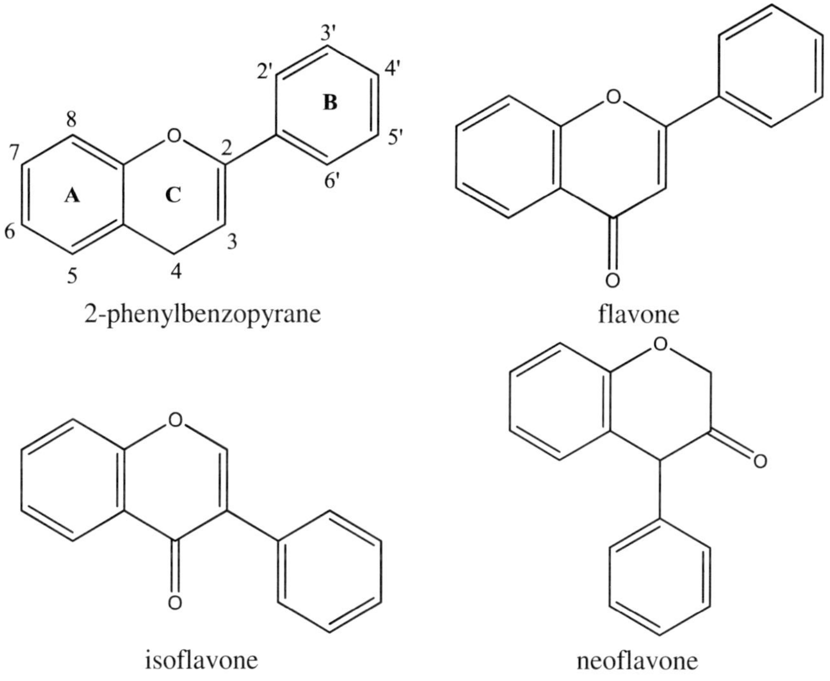


Figure 4. 2-phenylbenzopyrane, flavone, isoflavones and neoflavone structures.

Flavonoids are primarily found in plants. Some flavonoids are pigments but others are not. There are about 6,000 known compounds in the flavonoid category which in certain cases have shown to reduce the risk of cancer and heart disease, while others limit menopause symptoms or strengthen bones [26].

In the case of the peanut, flavonoids and derivatives can be found in both infected and uninfected *Arachis* plants, the most common being "eriodictyol, luteolin, chrysoeriol and dihydroquercetin".

However, flavonoid compounds such as isorhamnetin glucoside isolated from the water-soluble fraction of peanut skin, or quercetin isolated from the extracts of germinating peanut have been identified [26].

The isoflavones "bichanin A, daidzen, genistein and formonetin" can also be found in peanut oil [12]. The latter are expected to be recognized as essential nutrients [26].

## PHENOLIC ACIDS

Phenolic acids are known to act as antioxidants by producing hydrogen or electrons. In addition, their stable radical intermediates prevent the oxidation of various food ingredients, particularly fatty acids and oils [27].

In most cases, phenolic acids are not found in a free state, except at trace levels, but as combined forms. Nevertheless, some exceptional situations can cause phenolic acids to accumulate in a free form: brutal extraction conditions, physiological disturbances, contamination by microorganisms and anaerobiosis.

Furthermore, phenolic acids have been shown to be present in high concentration in the outer layers of plants such as the peel, shell, and hull, in order to protect inner materials such as the cotyledon [28].

From the extracts of germinating peanuts, phenolic acids vanillic, protocatechuic, ferulic and caffeic were isolated [12]. In addition vanillic acid, protocatechuic, ferulic, and *p*-coumaric (Figure 5) have been found in peanut roots inoculated with mycrorrhizal and Rhyzobium.

Ferulic and *p*-coumaric acids have been identified as major compounds from dry-roasted peanuts [28]. These phenolic acids are reported to have antioxidant effects [28].

In dry-roasted peanuts, *p*-coumaric acid seems to be responsible for the major antioxidant activity in the isolated polyphenols [28]. Moreover, phenolic acids such as caffeic, *p*-coumaric, ferulic and proto-catechuic acids are shown to exert an antifungal effect [27].

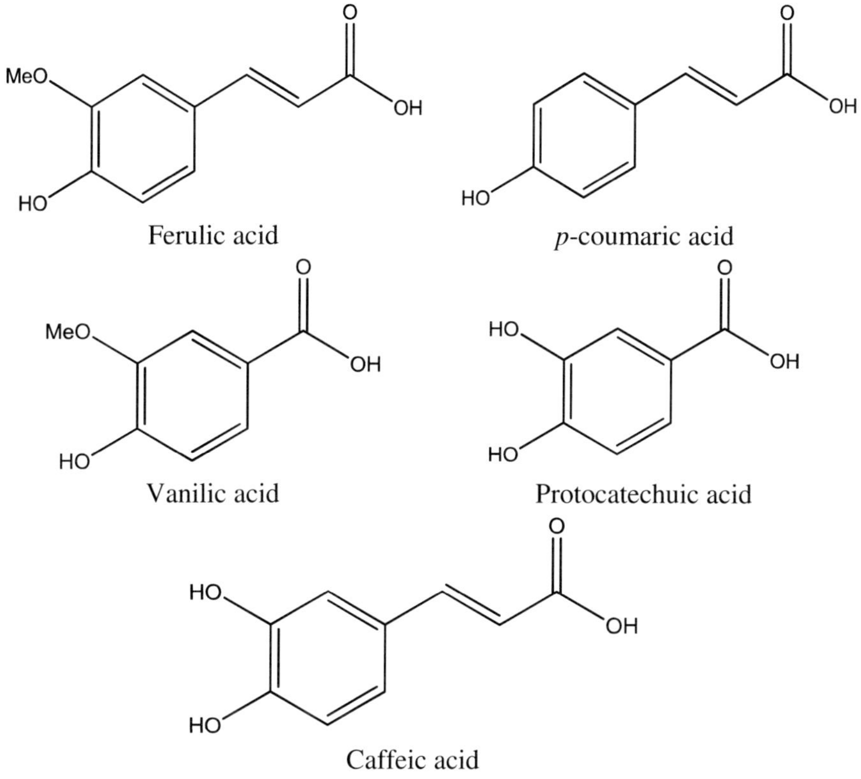


Figure 5. Ferulic, *p*-coumaric, Vanilic, Protocatechuic and Caffeic acids structures.

# PHYTOSTEROLS

Phytosterols (PS) are one of the most important bioactive constituents in peanuts. They are unsaponifiable triterpenes found in the lipid fraction, with a cyclopentanoperhydrophenanthrene ring structure (Figure 6), and are sometimes referred to as 4-desmethylsterols [29]. More than 250 phytosterols and related compounds have been described in the literature [30]. They are structurally similar to cholesterol but differ in the side chain at C24 and/ or in the position and configuration of double bonds. In foods, phytosterols can exist in four forms: free sterols, steryl esters of fatty or phenolic acids, steryl glucosides, and acylated steryl glucosides [30].

In peanut butter, peanut oil and groundseed *β*-sitosterol, campesterol, stigmasterol, α-spinasterol, Δ5-avenasterol and Δ7-avenasterol are found

(Figure 6) [12]. Other phytosterols found in *Arachis are* lophenol, 24-ethyllophenol, obtusifoliol (Figure 6), 31-norcyclartenol, cycloleucalenol, gramisterol and citrostadienol (Figure 6) [29].

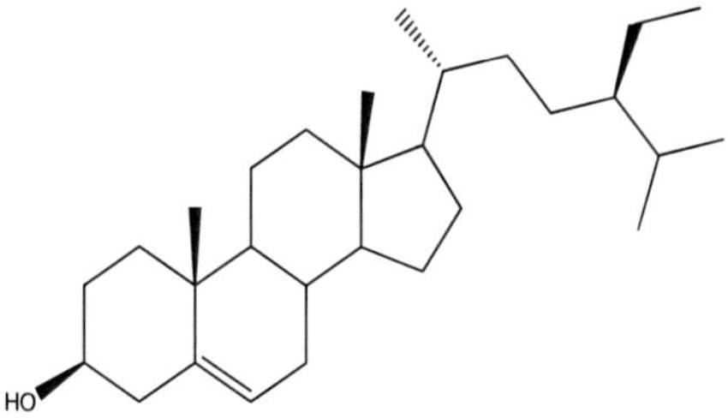


β-sitosterol

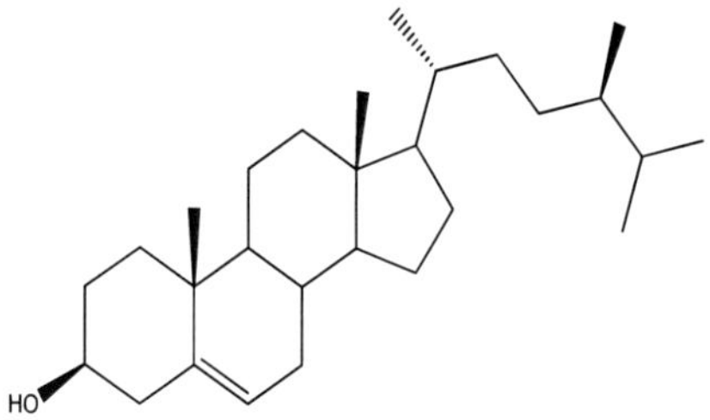


Campesterol

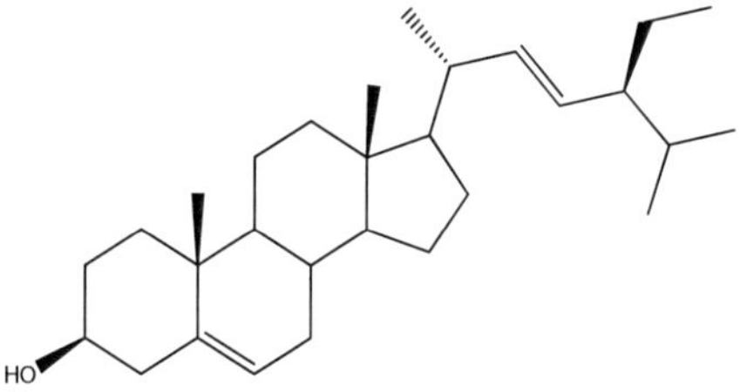


Stigmasterol

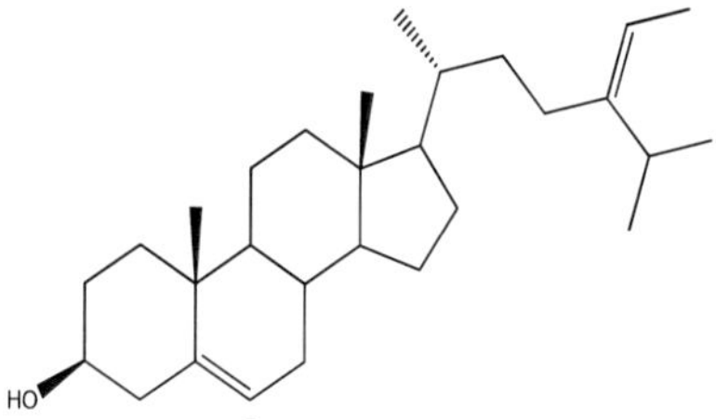


$\Delta^5$-avenasterol

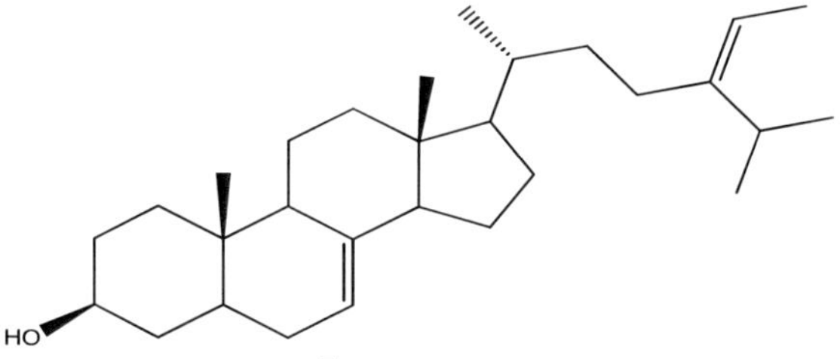


$\Delta^7$-avenasterol

Figure 6. Detailed chemical structures of β-sitosterol, Campesterol, Stigmasterol, $\Delta^5$-avenasteroland and $\Delta^7$-avenasterol.

The major sterol in peanut is the *β*-sitosterol. It reaches concentration values of 554.8 mg/100g of oil in the immature wild peanut while 267-309 mg/100 g of oil found in immature peanut cultivars [29]. At the maturity stage, the total 4-Desmethylsterols content is approximately 192-203 mg/100 of oil in cultivars and wild peanut [29]. Phytosterols in the peanut have several bioactive properties with possible benefits for human health [29]. They contribute to lowering serum cholesterol levels [29] and are also considered to have anti-inflammatory, anti-bacterial, anti-artherosclerotic, anti-oxidative, anti-ulcerative and anti-tumoral properties in humans [31-33], as well as contributing to the oxidative and thermal stability and shelf-life of vegetable oils [34]. Additionally, they are useful emulsifiers for cosmetic manufacturers and supply the majority of steroidal intermediates and precursors for the production of hormone pharmaceuticals [35]. Phytosterols might also prevent the development of colon cancer and benign prostatic hyperplasia [36]. Among different plant sterols, $\beta$-sitosterol has been intensively investigated and has been shown to exhibit anti-inflammatory; antineoplastic, antipyretic, and immunomodulating activities [37]. Also, $\beta$-sitosterol, campesterol and stigmasterol exert antioxidant effects.

## POLICOSANOL

Policosanol (PC) is the common name of a mixture of high molecular weight (HMW) aliphatic primary alcohols ($C_{20}$ to $C_{36}$), originally isolated from sugar cane (*Saccharum officinarum* L.) [38]. Peanut plants represent a good sources of PC with their health-enhancing components. PC is present in the fruits, leaves, surfaces of plants and whole seeds [38]. It contains mainly docosanol ($C_{22}$) (Figure 6), tetracosanol ($C_{24}$), hexacosanol ($C_{26}$), octacosanol ($C_{28}$) and triacontanol ($C_{30}$) [39]. In addition to saturated n-alcohols, the presence of unsaturated alcohol compounds such as (Z) Octadec-9-en-1-ol, Eicosén-1-ol, Heneicosén-1-ol, Docosén-1-ol (Figure 7), and Tetracosen-1-ol have been also reported [38].

The major PC components in peanut oil are successively: Hexadecan-1-ol, (Z) Octadec-9-en-1-ol, Heptadecan-1-ol, Octadecan-1-ol, and Docosan-1-ol. The total PC content of whole peanut samples varied from 11.18 and 54.19 mg/100g of oil and was higher than those of beeswax and whole sugar cane, which are sources of dietary supplements containing policosanol.

Policosanol has been shown to decrease platelet aggregation, endothelial damage, and foam cell formation. The effectiveness of policosanol as a lipid-lowering agent, in various different populations has been extensively studied [39]. PC is supposed to downregulate the cellular expression of hydroxymethylglutamyl coenzyme Z (HMG-CoA) reductase; the rate-limiting enzyme in cholesterol synthesis, [23].Furthermore, it also enhances LDL catabolism [40]. In addition, PC improves protection of lipoproteins against lipid peroxidation, both in the lipid and the protein moieties [41].

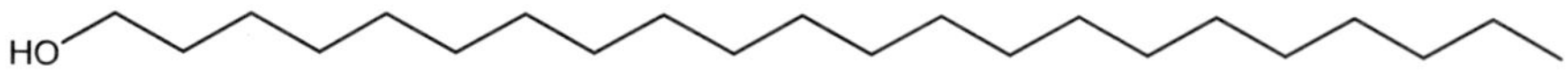

Figure 7. chemical structure of Docosan-1-ol.

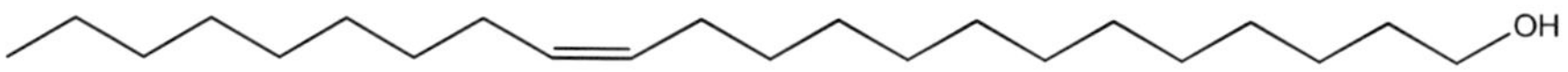

Figure 8. chemical structure of cis-13-docosen-1-ol.

## Squalene

Squalene (2,6,10,15,19,23-hexamethyl-2,6,10,14,18,20-tetracosahexaene) ($C_{30}H_{50}$), is a terpenoid hydrocarbon occurring in high concentrations in the peanut (Figure 9) . It is present in all plant oils and fats except flaxseed, grapeseed, and soybean oils [42].It is accompanied by *n*-alkanes in the range of $C_8$-$C_{35,}$ being more abundant in those between $C_{21}$ and $C_{35}$, in which alkanes with an odd number of carbon atoms predominated over those with even numbers [43].

In peanut butter, oil and groundseed, the squalene content is about 815.45 mg/100g of oil for immature peanut cultivars, while in the case of mature peanuts, the maximum level is about 346.74 mg/100g of oil [44].At these time-dates, wild and cultivar peanuts are considered as rich a source of squalene as olive and pumpkin oil.

Squalene is also regarded as one of the compounds responsible for the beneficial effects against certain type of cancers [42]. Its antioxidant activity can reinforce the endogenous antioxidant system against oxidative damage [42].Together with other components of the non-saponifiable fraction; it also prevents the autoxidation process and thus contributes to the stability of the oil, its aroma and its flavor characteristics [42].

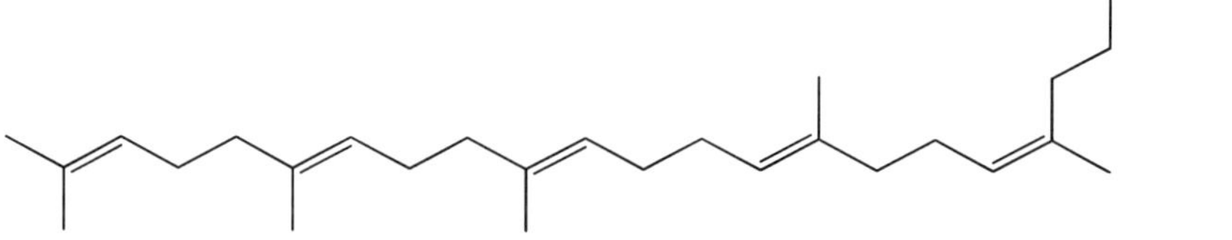

Figure 9. Squalene.

# Conclusion

The species *A. hypogaea* is commonly seen as a good source of oil and protein, but studies have shown that peanuts have a strong potential as bioactive food. The health-promoting effects of peanut consumption are attributed to their fatty acid profiles as well as to other beneficial functional constituents including phytosterols, policosanol, polyphenols and micronutrients like tocopherols. The numerous nutrients and bioactive compounds endogenous to peanuts together with their potential synergies truly make the peanut a desirable plant food. These bioactive compounds could successfully be used in peanut breeding situations to provide a good source of natural bioactive compounds that could be incorporated into functional foods and nutraceuticals. Further in vivo studies should be made in order to select the appropriate peanut cultivars in terms of nutritional and therapeutic profiles, and to enable the production of a functional peanut oil.

# References

[1] J. C. King, J. Blumberg, L. Ingwersen, M. Jenab and K. L. Tucker. *J. Nutr.* 138, 1736 (2007).

[2] H. Lou, Y.Yamazaki, T. Sasaki, M. Uchida, H. Tanaka and S. Oka. *Phytochem.* 51, 297 (1999).

[3] T. Vijaya, K. C. Maouli and S. D. Rao. *J. Global Pharma Technol.* 1, 4 (2009).

[4] S. Al-Qura'n. *J. Nat. Prod.* 1, 10 (2008).

[5] J. Higgs. *Nut. Food Sci.* 33, 56. (2003).

[6] P. M. Kris-Etherton, F. B. Hu and E. Ros, J. Sabate. *J. Nutr.* 138, 1746S. (2007).

[7] N.Okarter, C. S. Liu, M.E. Sorrells and E. H. Liu *Food.Chem.* 119, 249. (1999).

[8] V.S. Sobolev and R. J. Cole. *J. Agric. Food.Chem.* 47, 1435 (1999).

[9] A. Cassidy. *Proc. Nutr. Soc.* 55, 399-417. (1996).

[10] W. J. Lampe and J. L. Chang. *Seminars in Cancer Biol.*. 17,347 (2007).

[11] S.T. Talcott, S. Passeretti, C.E. Duncan and D. W. Gorbet. *Food Chem.* 90, 379 (2005).

[12] R.M. Lopes, T. Agostini-Costa, M. A. Gimenes and D. Silveira. *J. Agric. Food. Chem.*59,4321. (2011).

[13] B. Craft, A. Kosinska, R. Amarowiez and R.B. Pegg. *Plant foods Hum Nutr.* 65, 311. (2010).

[14] C. E. Duncan, D. W. Gorbet and S. T. Talcott. *Food Res. Int.* 39, 898. (2006).

[15] T.H. Sanders. *J. Agric. Food. Chem.*49, 2349. (2001).

[16] G. Horvath, L. Wessjohann, J. Bigirimana, M. Jansen, Y. Guisez, R. Caubergs et al. *Phtyochem.* 67, 1185. (2006).

[17] R. S. Jonnala, N. T. Dunford and K. E. J. Dashiell. *Food Compos. Anal.* 19, 601. (2006).

[18] U. Stervbo, O. Vang and C. Bonnesen. *Food Chem.* 101, 449. (2007).

[19] J. Chong, A. Poutaraud and P. Huguene. *Plant Sci.* 177, 143. (2009).

[20] A. Cassidy, B. Hanley and R. M. Lamuela Raventos, *J. Sci. Food Agric.* 80, 1044. (2000).

[21] V. S. Sobolev, B. Z. Guo, C. C. Holbrook and R. E. Lynch. *J. Agric. Food. Chem.* 55, 2195. (2007).

[22] J. S. Kim, S. Y. Lee and S. U. Park, *Afr. J. Biotechnol.* 7, 3788. (2008).

[23] Y. Chukwumah, L. Walker, B. Vogler and M. Verghese. *J. Agric. Food. Chem.*55, 9266. (2007).

[24] L.L. Lin, C.Y. Lien, Y.C. Chen and K.L. Ku. *J. Chromatogr. B. Anal. Technol. Biomed. Life Sci.* 853, 175 . (2007).

[25] Z. Lu, B. Cheng, Y. Hu, Y. Zhang and G. Zou. *Food Chem.* 17 (2009).

[26] J. H. Lee, I. Y. Baek, N. S. Kang, J. M. Ko, H. T. Kim, C. S. Jung et al. *Food Sci. Biotechnol.* 16, 29–36 (2007).

[27] S.R. Pradeep and M. Guha. *Food.Chem.* 126,1643. (1999).

[28] C. E. Duncan, D. W. Gorbet and S. T. Talcott. *Food Res. Int.* 39, 898. (2006).

[29] A.O. Cherif, H. Trabelsi, M. Ben Messaouda, B. Kâabi, I. Pellerin et al. *J Agric Food Chem.* 58, 8709. (2010a).

[30] E. C. Shin, R. B. Pegg, R. D. Phillips and R. R. Ettenmiller. *J Agric Food Chem.* 58,9137. (2010).

[31] M.J. Lagarda, G. Garcia-Liatas and R. J. Farre, *Pharm. Biomed. Anal.* 41, 1486. (2006).

[32] R.A.; Moreau, B.D. Whitaker and K.B. Hicks. *Prog. Lipid Res.* 41, 457. (2002).

[33] T.H.J.; Beveridge, T.S.C. Li, and J.C.G Drover. *J. Agric. Food Chem.,* 50, 744. (2002).

[34] R. Przybylski and N.A.M. Eskin. *Inform.* 17, 187. (2006).

[35] C. Grunnwald. *Annu Rev Plant Physiol.* 26, 209. (1975).

[36] A.B. Awad, K.C. Chan, A.C. Downie and C.S. Fink. *Nutr Cancer.* 36, 238. (2000).

[37] M. Careri, L. Elviri and A. Mangia. *J. Chromatogr. A.* 935, 249. (2006).

[38] A. O. Cherif, M. Ben Messaouda, B. Kaabi, S. Boukhchina, C. Pepe and H. Kallel, *J.Agric. Food Chem.* 58, 12143. (2010).

[39] S. Irmak, N. T. Dunford and J. Milligan. *Food Chem.* 95, 312. (2006).

[40] R. Menéndez, S.I. Fernandez, A. Del Rio, R.M. Gonzalez, V. Fraga, A.M. Amor et al. *Biol Res.* 27, 199. (1994).

[41] R. Menéndez, V. Fraga, A. M. Amor, R. M. González and R. Mas. Physiol. Behav. 67, 1. (1999).

[42] C. I.G. Tuberoso, A. Kowalczyk, E. Sarritzu and P. Cabras. *Food. Chem.* 103, 1494. (2007).

[43] A. Cert, W. Moreda and M. C. Pérez-Camino. *J. Chromatogr. A.* 881, 131. (2000).

[44] A.O. Cherif (PhD Thesis). (2012).

In: Peanuts: Bioactivities and Allergies ISBN: 978-1-62417-387-5
Editors: I.L. Expósito, A.B. Blázquez 

*Chapter 4*

# PHENOLIC COMPOUNDS: FUNCTIONAL COMPONENTS IN PEANUTS

***E. Vázquez Rodríguez*[*] *and I. López Expósito***
Institute of Food Science Research CIAL (CSIC-UAM),
Madrid, Spain

## ABSTRACT

Natural products from plants have been used for centuries for the prevention and treatment of diseases in humans. Nowadays, the beneficial compounds present in these natural products are being used by the food industry in functional foods and have been called by various terms such as phytochemicals, functional and bioactive components.

One of these phytochemicals found in peanuts are phenolic compounds. There are numerous scientific studies demonstrating their potential effects on human health since they may act as antioxidants, preventing cardiovascular diseases or inhibiting carcinogenesis.

This chapter presents the composition and evidence about the bioactive properties of these compounds, showing peanuts as an interesting source of healthy ingredients.

---

[*] E-mail address: erika.vazquez@uam.es.

# Introduction

Natural products from plants have been used for centuries for the prevention and treatment of certain illnesses. Hippocrates said: "Let your food be your medicine", claiming the importance of food as a source of health and its properties on some conditions. Thanks to the scientific development and the appearance of new analytical techniques, as well as the evolution of molecular biology, some questions regarding the bioactive properties of food have been clarified. In this way, specific compounds contributing to those properties, together with their metabolism and biological activity have been established.

Nowadays, beneficial compounds in natural products are being used by the food industry to generate functional foods. They have been called phytochemical elements, functional compounds and bioactive compounds. Over the last few years, a dramatic increase of general interest in them has taken place and, more precisely, in phenolic compounds, because of their potential role in human health. Scientific research has shown the effects of these substances as they may act as antioxidants [1-3], blocking the activity of bacterial and viral toxins [4-8], decreasing cholesterol absorption [9,10], preventing from cardiovascular diseases [11-14], or inhibiting carcinogenesis [15-18]. Phenolic compounds are present in the majority of vegetables, such as fruits, pulses, nuts, tubers, cereals and aromatic herbs. However, in many cases these compounds are located in non-edible parts of the plants, creating a way to obtain them from the subproducts of the food industry for their later use as additives or in formulating functional or nutraceutic foods.

Peanuts have been considered an interesting functional food recently. In the past, peanuts were considered unhealthy due to their high fat content, close to 50%. However, that opinion has changed, as it has been shown that peanuts contain many beneficial components for human health, such as vitamins, polyunsaturated fatty acids and sterols [19, 20]. A big amount of functional compounds have been isolated, identified and quantified (Table 1) and their functional activities on human health have been studied. Thus, in the last decades many clinic and epidemiological studies have related peanuts consumption with cardiovascular health [23, 24]. Furthermore, recent research also relates peanuts with antioxidant [25], antibacterial, anti-inflammatory and antitumoral properties [20, 26, 27]. Poliphenols are one of the main bioactive components found in peanuts.

**Table 1. Beneficial components present in peanuts [20, 21, 22]**

| Component | Amount | Unit |
|---|---|---|
| Dietary fiber | 8.00 | g[1] |
| Total monounsaturated fatty acid | 24.64 | g[1] |
| Total polyunsaturated fatty acid | 15.69 | g[1] |
| Vitamin E ($\alpha$-tocopherol) | 6.93 | mg[1] |
| Folate | 145.00 | µg[1] |
| Magnesium | 176.00 | mg[1] |
| Potassium | 658.00 | mg[1] |
| Total phytosterols (raw, shelled peanut) | 55.1-126.9 | mg[1] |
| $\beta$- sitosterol | 47.20 | mg[1] |
| Campesterol | 5.80 | mg[1] |
| Stigmasterol | 7.70 | mg[1] |
| Total polyphenols (raw peanut) | 25.71 | mg GAE[2] |
| Total polyphenols (raw with skin) | 28.71 | mg GAE[2] |
| Total proanthocyanidins | 15.60 | mg[1] |
| Procyanidins dimmer (peanut skin) | 111.3 | mg[3] |
| Procyanidins trimer (peanut skin) | 221.3 | mg[3] |
| Procyanidins tetramer (peanut skin) | 296.1 | mg[3] |
| Catechins (peanut skin) | 16.1 | mg[3] |
| (-)-Epigallocatechin | 0.66 | mg[1] |
| p-Coumaric acid | 8.82 | mg[1] |
| trans-Resveratrol | 5.50 | µg[1] |
| Daidzein | 0.13 | mg[1] |
| Genistein | 0.024 | mg[1] |
| Biochanin A | 0.043 | mg[1] |
| Formononetin | 6.8 | µg[1] |

[1] content expresed per 100 g peanut.

[2] content expresed as gallic acid equivalent per g peanut (mg GAE/g).

[3] content expresed per 100 g peanut skin.

## PHENOLIC COMPOUNDS

They are compounds responsible for colour, taste and astringent properties of food, as well as the prevention of lipooxidation and protection of vitamins and enzymes. The appearance of phenolic compounds in food is changeable regarding both quality and quantity. These differences exist as a consequence of genetic varieties, environmental conditions and growing stages.

Moreover, differences in concentration and composition may also take place in the different parts of the fruit. Phenolic compounds come from the secondary metabolism of vegetables and are found in small amount in most vegetable tissues both in monomers (acylated or glycosylated), polymer or complex forms (along with other phytomolecules). In peanuts, they are mainly in the external parts of the fruit, such as peels or rinds, although they may also be found in the grain and roots [20].

Phenolic compounds are synthesized from a little part of the carbohydrates created during the photosynthesis [28, 29]. The byosynthesis process is summarized in Figure 1.

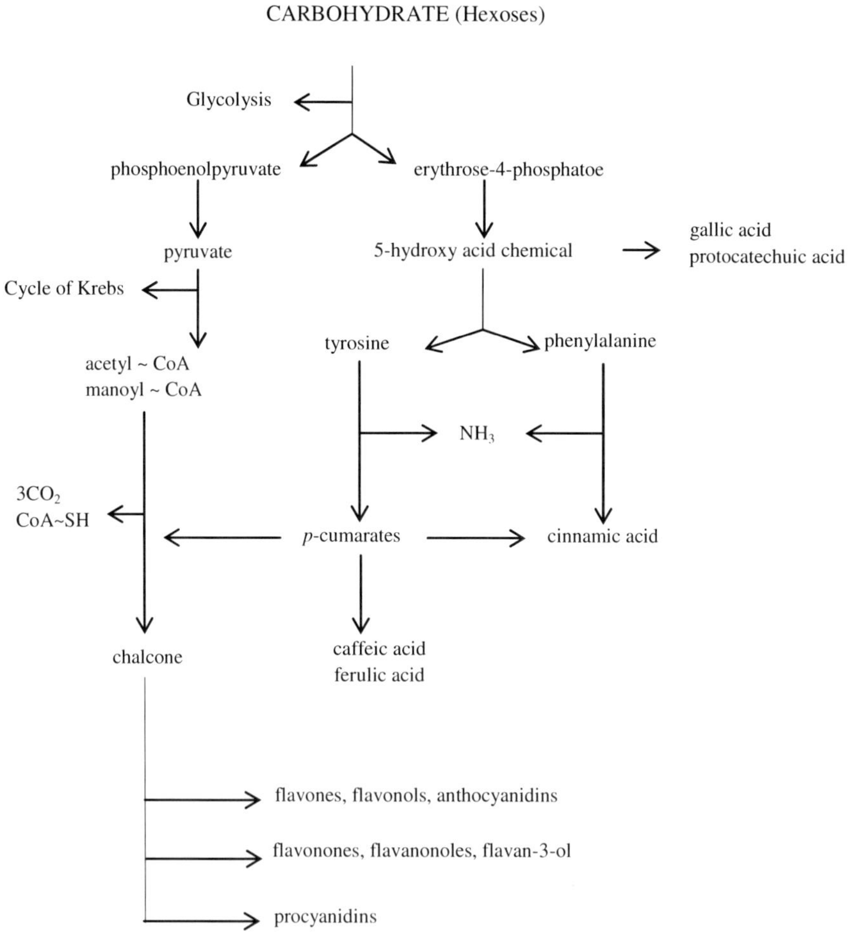


Figure 1. General outline of the biosynthesis of the phenolic compound.

Phenolic compounds are a group of heterogeneous substances, characterized by having an aromatic ring with, at least, one hydroxyl group and a lateral functional chain. Their classification is approached regarding its carbon skeleton. They may be divided into two main groups: flavonoids and non flavonoids [2, 30, 31].

Flavonoids form a group of phenolic compounds identified by having a common structure $C_{15}$ (C6-C3-C6) [32]. In this structure, the heterocyclic is called ring "C", the ring bound to it is called "A" and the phenolic constituent is known as "B" (Figure 2).

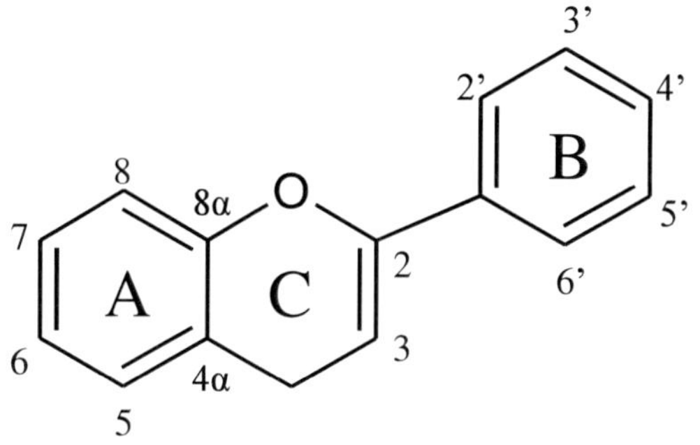


Figure 2. Chemical structure of flavonoids.

Based on the saturation and on the oxidation stage of ring C, they can be divided into the following basic structures: flavanone, flavones, flavonols, isoflavones, flavanols and anthocyanidins [20, 33, 34].

Flavanols or flavan-3-ol constitutes the main phenolic compounds present in peanuts. They are divided into two main groups: catechins and proanthocyanidins. Catechins are characterized by having 2 asymmetric carbon in positions 2 and 3, obtaining 4 isomers, the (+)/(-)-catechin and the (+)/(-)-epicatechin. On the other hand, glycosylated and acylated derived forms may be found.

The most common of the last ones are the result of the replacement of group OH from position 3 of the heterocycle by a molecule of gallic acid, giving 3-O-gallates (+)/(-) (gallo) catechin (and (+)/(-) epi (gallo) catechin (EGC) as a result. These compounds are generally extracted from leaves, roots, barks and vegetable seeds [35, 36] being their great variety a result from capacity of the catechins to create polymers of a different stage of polymerization, as well as of a different stage of branching. Catechin polymers are called proanthocyanidins.

## Composition and Variation

The content of phenolic compounds in food is highly variable in both quantity and quality. This content may vary depending on the family and species [33].

Francisco et al. [37] claimed that the total content of polyphenols in peanut skin varies depending on the type analysed, being the rank of the total content of polyphenols of the different extracts of 101.43-280.42 mg/g for Runner variety, 106.60–148.84 mg/g for Virginia and 95.56–136.80 mg/g for the Spanish one. This variability may be due to genetic differences, environmental conditions and the growing or degree of maturation of the plant or fruit [20, 33]).

Consequently, differences in the composition of some phenolic compounds may be noticed depending on the season of the year in which the harvest is fulfilled. As an example, Chen et al. [38] noticed that resveratrol content in the roots of peanut was higher in the summer rather than in the spring.

Inside one single vegetable, the main variability may be noticed in the different parts of the plant. While the simplest phenolic compounds tend to concentrate on the fruit, the complex or polymeric forms are in the hard parts of the plant.

The main phenolic compounds isolated in peanut are showed in Table 2 and they follow this tendency, in which the non flavonoids compounds such as the phenolic acids and stylbens are concentrated in the grain, while the flavonoids compounds are mainly in the peel.

These contents may have variations throughout the development of the plant and fruit. Daigle et al. [39] found out that when peanut is unripe the main flavonoid in the shells was the eriodictyiol, whereas when the maturation was making progress the main compound was the luteolin.

Some research has shown that the concentration of the total phenolic compounds and luteolin increases with the maturity of the shell [40]. Sovolev et al. [41] in their studies on the content of trans-resveratrol in peanuts claim that unripe peanuts grains are more capable of synthesizing this compound than ripened grains. But once they are picked, the processing as well as the storage conditions may cause changes in their composition.

In this way, it has been noticed that during the processing and storage of food some epimerization, degradation and de-polymerization of flavanol oligomers and polymers [34] may take place, as well as hydrolysis of glycosylated flavonoids and aglycones decomposition [42]. It has also been observed in some cases an increase of the total phenolic compounds content. This increase may happen due to the development of Maillard products and/or the release of phenolic compounds, mainly when roasting temperatures are reached [25,37]. Yu et al. [43] did some research about the effect of processing on the phenolic compounds content of the rind of peanut and reported that after toasting there was an increase of 39.5%.

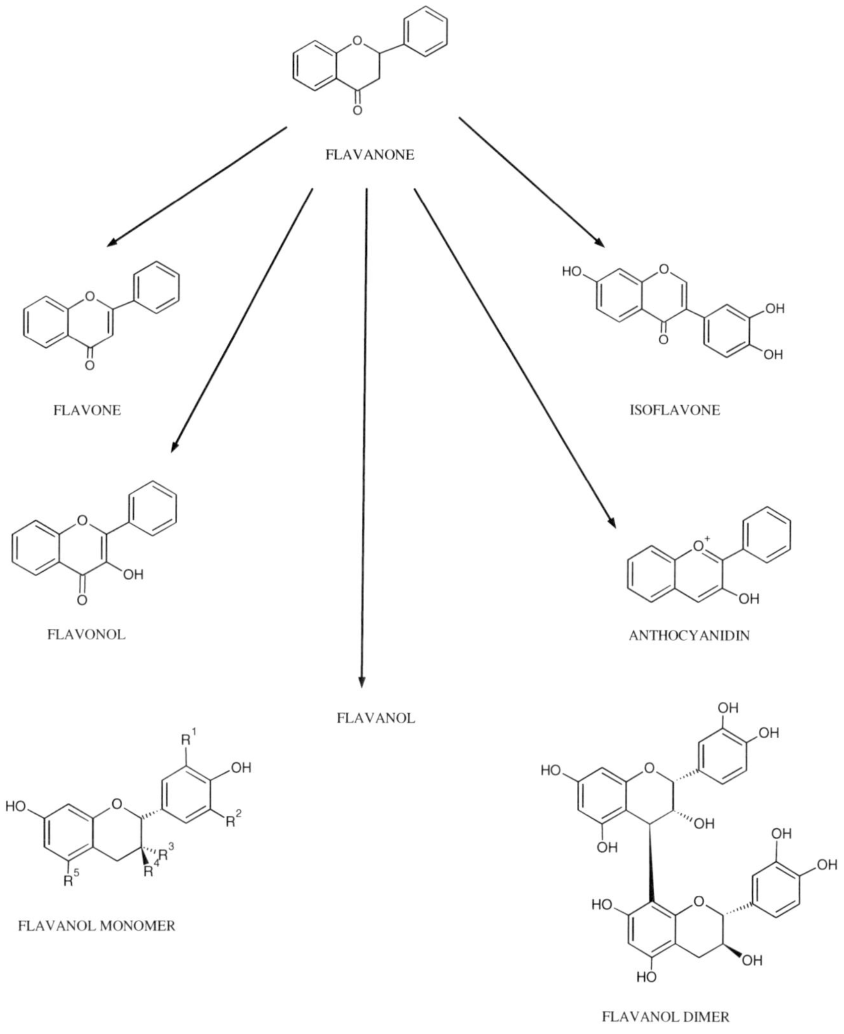


Figure 3. Structure and origin of the main groups of flavonoids [33].

**Table 2. Polyphenols isolated from peanut**

| Raw material | Compounds | Reference |
|---|---|---|
| Peanut kernels | resveratrol | 41 |
| | dihidroquercetin | 20 |
| | p-hidroxybenzoic | 45 |
| | p-coumaric acid | 20 |
| | Vanillin | 46 |
| | biochanin A | 19, 44 |
| | daidzein | |
| | genistein | |
| Germinated peanut | vanillic acid | 19 |
| | protocatechuic acid | |
| | ferulic acid | |
| | caffeic acid | |
| | quercetin | |
| Peanut hulls | luteolin | 19 |
| | eriodictyol | |
| | 8-isopentenyl-luteolin | |
| | dihydroquercetin | |
| | chrysoeriol | |
| | 8-isopentenyl-chrysoeriol | |
| | 5,7-dihydrixychromona | 20 |
| | vanillin | 46 |
| Peanut flour | p-hidroxybenzoic | 47 |
| | trans-p- coumaric acid | |
| | trans-ferulic acid | |
| | trans-caffeic acid | |
| | trans-sinaptic acid | |
| Peanut roots | resveratrol | 19, 38 |
| | pterostilbeno | 19 |
| | vanillic acid | |
| | protocatechuic acid | |
| | ferulic acid | |
| | caffeic acid | |
| | chlorogenic acid | |
| Peanut leaves | resveratrol | 20 |
| | formonetin | 48,49 |
| | medicarpin | |
| | memethylmedicarpin | |
| | isomedicarpin | 50 |
| | chlorogenic acid | 19 |
| | ceochlorogenic acid | |
| | 1-caffeoyl-4-deoxyquinnic acid | |
| Peanut skin | Chlorogenic acid | 43 |
| | Caffeic acid | |
| | Coumaric acid | |

| Raw material | Compounds | Reference |
|---|---|---|
| Peanut skin | Ferulic acid | |
| | Ethyl protocatechuate | 51 |
| | Isorhamnetin glucoside | 19 |
| | Isorhamnetin 3-O-[2-O-β-glucopyranosyl-6-O-α-rhamnopyranosyl]-β-glucopyranoside | 52 |
| | Isorhamnetin 3-O-[2-O-β-glucopyranosyl-6-O-α-rhamnopyranosyl]-β-glucopyranoside | |
| | trans-resveratrol | 20, 43 |
| | (+)-catechin | 42 |
| | (-)-epicatechin | 42, 43 |
| | Epigallocatechin | 43 |
| | Catechin gallato | |
| | Epicatechin-(4β→8)-catechin | 20 |
| | Epicatechin-(2β→O→7,4β→8)-catechin | |
| | Epicatechin-(2β→O→7,4β→8)-epicatechin | |
| | Epicatechin-(2β→O→7,4β→6)-catechin | |
| | Epicatechin-(2β→O→7,4β→6)-ent-catechin | |
| | Epicatechin-(2β→O→7,4β→6)-ent-epicatechin | |
| | Epicatechin-(2β→O→7,4β→8)-ent-epicatechin | |
| | Proanthocyanidin $A_1$ | 42 |
| | Proanthocyanidin $A_2$ | |
| | Procyanidin dimer $A_1$ | 22, 42, 53-55 |
| | Procyanidin dimer $A_2$ | |
| | Procyanidin dimer $B_1$ | 42, 55, 56 |
| | Procyanidin dimer $B_2$ | |
| | Procyanidin dimer $B_3$ | |
| | Procyanidin dimer $B_4$ | 22, 42, 56, 64 |
| | Procyanidin A dimmers, trimers, tetramers | 53, 54, 57 |
| | Procyanidin B dimmers, trimers, tetramers | |
| | Proanthocyanidins petamers-dodecamers | 42, 56 |

However, different heat treatments do not modify the content of total phenolic compounds similarly. Yu et al. [22] studied the effect of three different processes of peanut peeling on the content of total phenolic compounds, showing that peeling by toasting has a moderate effect on the concentration of these compounds, but peeling by whitening causes a loss of 89%, obtaining the values of 130.8, 15.1 and 124.3 (mg GAE/g dry skin) for the extracts obtained from peanuts peeled directly, through whitening and through toasting, respectively.

Sovolev et al. [41] did a comparative study the content of resveratrol of different peanuts and peanuts commercial products concluding that the biggest resveratrol content is in boiled peanuts and the smallest in toasted ones, being these values of 5.138±2.849 μg/g and 0.055±0.023 μg/g, respectively. Chukwumah et al. [44] in his study about the effects of different treatments in the content and polyphenolic profile of peanuts, showed that processing did not affect the flavonoids content, whereas it did have an influence on the content of the total phenolic compounds.

The biggest difference in the concentration of these compounds was noticed between raw and boiled peanuts, on one hand, and toasted ones, on the other, being the raw and boiled peanut wherein the content of phenolic compounds and flavonoids was bigger.

Moreover, a significantly higher content of flavonoids and total phenolic compounds in boiled peanuts facing the raw ones was found. This may happen because boiled peanuts are processed with their rind, which has a high content in polyphenols. Due to that, during the boiling grains may absorb the water enriched in hydrosoluble polyphenols.

## BIOAVAILABILITY

Polyphenols have many beneficial properties; however, bioavailability and therefore the bioefficacy have to be taken into account. In this way, the polyphenol in the biggest concentration in food does not have to be the one with the biggest biological activity, because its metabolism may be weak and the metabolites derived might be in a low concentration in the organism. Furthermore, polyphenols that have good results *in vitro*, may have a weak activity *in vivo,* because their metabolism by the digestive system and the liver may generate metabolites different from the native compounds. After ingestion of 50 mg of polyphenols (equivalents of aglycone) the concentration rank in total plasma metabolites is between 0-4 μg/L and the rank of relative urinary excretion 0.3-43% of the dose taken [58]. Moreover, their chemical structure, glycosylation, functional groups and grade of polymerization condition the bioavailability of these compounds. After many studies about metabolism and polyphenol bioavailability it has been shown that the gallic acid along with the isoflavones are the polyphenols which are better absorbed, followed by the catechins, flavanones and quercetin glycosides. On the other hand, the polyphenols which have shown the worst absorption are the proanthocyanidins due to their grade of polymerization, and the galloyl catechins, as the grade of galloylation also affects the effectiveness of the absorption in these compounds. In Table 3 some metabolic parameters of some of the main phenolic compounds in peanuts are shown. The fact of finding certain concentrations of these compounds or their metabolites in the blood indicates that they may be metabolized and able to go through the intestinal barrier. Albeit not all the phenolic compounds are metabolized, absorbed and excreted similary [34, 59, 60].

After performing bioavailability studies in humans, there has been seen a variability in the products found after the ingestion of phenolic compounds, so it must be taken into account that the metabolism and bioavailability of these compounds depends also on the individual.

As it is shown in Table 3, quercetin glycosides are absorbed more efficiently than pure quercetin, demonstrating that glycosidation affects directly to the absorption of the polyphenol. This effect has also been observed with *trans*-resveratrol and its glycosylated form isolated from peanut [61]. Similarly, it can also be observed how the galloilation of catechins leads to a decrease in their absorption. Only free forms of epigallocatechin gallate have been found in blood plasma (77-90%). The low percentage of recuperation of the galloyl forms is not due to a process of de-galloylation, as this is a minor reaction in humans, but to its excretion in the bile [62].

The bioavailability of some of the hydroxycinnamic acids present in the peanut such as the caffeic, ferulic and chlorogenic acid has also been studied. Regarding the ferulic acid, identified in peel, roots and peanut sprouts, it has been shown that it is more easily absorbed in free form, however in some food, such as cereals it is esterified to arabinoxylans, making its absorption more difficult [58].

Luteolin is a flavone presents in the peanut rind. This compound has been given to volunteers and some research has revealed that it is metabolized producing a monoglucuronide, detected in plasma with some small amounts of luteolin unconjugated [1].

Other phenolic compounds found in peanuts are the isoflavones daidzein and genistein. The metabolism of both of them have been studied, both aglycones and glycosylated conjugates, and it has been proved that aglycones of these compounds are better absorbed than their glycosylated forms. Moreover, based on *in vitro* studies, it has been suggested that genistein may become fatty acid esters in plasma, making its incorporation to the LDL easier. Finally, another compound widely found in the peanut is resveratrol. This compound suffers sulfation and conjugation reactions with glucuronic acid and it is not found in plasma in free form [1].

**Table 3. Kinetic parameters from bioavailability [58]**

| | Mean $C_{max}$ (μmol/L plasm) | Mean Relative Urinary excretion (% of intake) |
|---|---|---|
| Gallic acid | 4.00 ± 0.57 | 37.7 ± 1.0 |
| Genistein | 2.56 ± 1.00 | 8.6 |
| Daidzein | 1.57 ± 0.52 | 27.5 |
| Quercetin glycosides | 1.46 ± 0.45 | 20.5 ± 1.2 |
| Epigallocatechin | 1.10 ± 0.40 | 11.1 ± 3.5 |
| Caffeic acid | 0.96 ± 0.26 | 10.7 |
| Epicatechin | 0.40 ± 0.09 | 18.5 ± 5.7 |
| Epigallocatechin gallate | 0.12 ± 0.03 | 0.06 ± 0.03 |
| Pure quercetin | 0.29 | — |
| Chlorogenic acid | 0.26 | 0.3 |
| Ferulic acid | 0.03 | 27.6 ± 17.6 |
| Proanthocyanidin dimers | 0.02 ± 0.01 | — |

All data correspond to a supply of 50 mg aglycone equivalent.
$C_{max}$, maximal plasma concentration.

# EFFECTS OF PHENOLIC COMPOUNDS ON HUMAN HEALTH

## Antioxidant Activity

One of the most studied properties of the phenolic compounds is their antioxidant activity [63, 64]. This activity is performed through several mechanisms; reacting with the free radicals, through the chelation of transition metals and through the inhibition and inactivation of enzymes. One of the main reactions in which the antioxidants are involved is the inhibition of lipid oxidation [36, 65], which does not only affect food quality, but also the organism at a cellular level.

Research by Nepote et al. [66] showed that the antioxidants present on the peel of toasted peanuts are equally effective than the addition of artificial antioxidants like BHT for their preserving during the storage and to protect them against lipid oxidation. Other works revealed that phenolic compounds present in peanuts such as quercetin, ferulic acid, catechin and epicatechin have good antioxidant properties in the DPPH and ABTS tests [67, 68]. Antioxidant activities from extracts of peanuts peel have been also performed through other methods [37, 42, 69]. Results of these studies allowed the comparison between the antioxidant potency of peanut extracts versus other foods. ORAC values of the peel extracts from peanuts, hazelnuts and almonds were 2.13, 3.05 and 1.07 mmol of TE/g, respectively [42]. However, ORAC values for EtOH, MeOH and water extracts performed under very good conditions of extraction were 2049.7, 2149.0 and 612.0 µmol of TE/g for dry skins, respectively [69]; compared with values of 92.56 and 92.09 µmol of TE/g for cranberries and blueberries and 42.34 µmol of TE/g for the Red Delicious apple, being the last ones hydrophilic extracts performed with acetone/water/acetic acid (70:29.5:0.5) [70].

In relation to the influence of the different thermal treatments which peanut suffers during processing, Craft et al. [25] performed a study of the antioxidant activity of processed peanuts through dry-roasted and oil-roasted compared with raw peanuts and determined that both the content in total phenolic compounds and the antioxidant activity increased in this way: raw < dry-roasted < oil roasted. They also observed an increase in *p*-cumaric acid content. Regarding the compounds formed during peanut roasting, it was shown that the pyrazines formed did not have antioxidant properties; however other compounds derived from the Maillard reaction, such as pyrroles and furans did have antioxidant activity [71]. Therefore it was hypothesized that, the moderate increase of the antioxidant activity after toasting could be because of the formation of these compounds.

## Protective Action of the Vascular System

It has been shown that phenolic compounds could protect the vascular system due to their antioxidant [13, 20], antithrombogenic and anti-inflammatory properties [37, 72].

Phenolic compounds like epi-gallo catechin gallate (EGCG), resveratrol, genistein, luteolin and procyanidins (trimers, tetramers and polymers and their gallates) have shown that they are able to modulate the production of nitric oxide (NO), an important vasodilator factor that also prevents from the platelet aggregation [1, 73-75]. Also, it has been shown that flavanones inhibit the formation of Angiotensin II by the angiotensin-converting enzyme I, in HUVEC cell cultures [76]. Therefore, phenolic compounds may decrease blood pressure through the modulation of vasodilators and vasoconstrictors. Appart from platelet aggregation, another important factor in the atherosclerotic plaque formation is the migration of macrophages and endothelial cells to the damaged area. Both *in vitro* and *in vivo* studies have proven the capacity of resveratrol, catechin, quercetin, luteolin and oligomeric procyanidins to block the platelet aggregation by inhibiting platelet adhesion to collagen [19, 20, 77-79]. On the other hand, many studies support that some phenolic compounds such as quercetin, EGC, EGCG, and resveratrol inhibit the proliferation and migration of vascular smooth muscle cells [1]. In a study, LPS-activated macrophages were treated with an extract of germinated peanut with a high content in *trans*-resveratrol, *trans*-arachidin-1, *trans*-arachidin-3 and *trans*-isopentenyl resveratrol. They found a significant inhibition in the production of prostaglandin $E_2$ ($PGE_2$) and NO, suggesting that the anti-inflammatory activity of these phenolic compounds was determined by the number and position of the hydroxyl groups and by the covalent binding of isopentenyl or isopentadienyl [27].

Other causes of atherosclerosis are hypercholesterolemia, obesity and diabetes. In this sense, many studies support the capacity of polyphenols to decrease the levels of cholesterol, as well as some parameters associated with obesity and secretion of insulin [77, 80-82]. Peanut's polyphenols have been studied in obesity by Moreno et al. [83]. They worked with two groups of rats fed with a fat-rich diet including peanut rind extract in one of them. They demonstrated an inhibition in pancreatic lipase and lipoprotein lipase enzymes in rats fed with the peanut rind extract this was accompanied a decrease in body weight, liver size, triglycerides in the liver and glucose and insulin in the serum.

## Phenolic Compounds for Cancer Prevention

Phenolic compounds may act at different levels to prevent cancer: as antioxidants, through anti-inflammatory mechanisms [15], through the regulation of expression of proliferative genes [16], inhibiting angiogenesis and cell invasion [17, 18, 84], paralyzing the cellular cycle and promoting apoptosis [16, 85, 86] or modulating the activity of enzymes involved in the detoxification among others.

Regarding their antioxidant activity, it has been demonstrated that these compounds may act inhibiting the production of reactive oxygen species by neutralizing the free radical, increasing the production of antioxidant enzymes or inhibiting the synthesis of prooxidant enzymes. It has also been shown that they have the capacity to reduce the DNA damage caused by different mutagens [87]. In this line Yen and Dhu [88] evaluated the antimutagenic power of peanut hulls extract facing different mutagens in Salmonella cultures. They saw that the antimutagenic capacity of the extract was dose-dependent and that also was dependant on the mutagen used.

Also, it has been shown that the phenolic compounds may act against a wide variety of cancers and that a single compound can perform its action on different cell lines. This is the case of resveratrol, one of the phenolic compounds found in peanut, which may act as an antiproliferative agent by inhibiting the expression of antiapoptotic proteins through different pathways in non-melanoma skin cancer, breast cancer, lung cancer, esophagus, colorectal, prostate, liver and pancreas cancer, and also facing fibrosarcoma, neuroblastoma and leukaemia; as well as inhibiting the replication of *Helicobacter pylori*, one of primary etiological determinants for gastric cancer [89].

### Protective Action for Neurodegenerative Diseases

In general, it has been found a protective relationship between the consumption of phenolic compounds and the risk of suffering neurodegenerative diseases [90-93]. The results of a study where 1640 persons were investigated throughout 5-10 years in two areas of the southwest of France, suggest that the consumption of 14.3± 5.85 mg/d of flavonoids in the diet reduces the cognitive decrease associated with neurodegenerative diseases in people at an age of 65 or over [94]. In general it has been postulated that the phenolic compounds may face these diseases by applying their antioxidant and anti-inflammatory properties [90, 92, 95-97]. However, it has also been observed that the phenolic compounds may act in a more precise way on some key factors of these diseases. For instance, the EGCG is able to regulate the proteolytic process of the Amiloyd Precursor Protein, and in fact, it has been checked that this compound reduces the generation of amyloid β peptide in Alzheimer disease [90, 92, 93, 98]. The same effect has been observed for resveratrol [99, 100]. On the other hand, phenolic compounds may act as neuroprotective factors facing Parkinson disease by blocking the activity of the dopamine transporter, which avoids the entrance of neurotoxins and increasing the bioavailability of the levodopa, commonly used for the treatment of Parkinson disease [95, 97].

## Conclusion

In the last decades many investigations and epidemiological studies relate consumption of food and drinks with a high content of polyphenols with the prevention of certain diseases. Cancer, cardiovascular and neurodegenerative diseases have the oxidative stress in common as an important part of their origin. In this way, and due to their antioxidant and modulation properties, phenolic compounds may take part in the prevention of these diseases. Because of their high content in phenolics compounds, peanuts are presented as a food with many beneficial properties for the organism. Their non-edible parts (skin, hulls, roots) which generally are part of the subproducts of the processing of peanut in the industry, are regarded now a source for obtaining these compounds, which may have an important application in the food industry as ingredients in the formulation of functional foods.

## REFERENCES


[1] Ø. M. Andersen and K. R. Markham. *Flavonoids. Chemistry, Biochemistry and Application.* Taylor & Francis Group, Boca Raton, FL, USA (2006).

[2] E. Grotewold. *The Science of Flavonoids.* Springer Science_Business Media, Inc., New York (2006).

[3] M. Ali Al-farsi and C. Yong Lee. Crit. Rev. Food Sci. 48, 877 (2008).

[4] L. C. Chiang, W. Chiang, M. Y. Chang, L. T. Ng and C. C. *Lin. Antivir. Res.* 55, 53 (2002).

[5] M. J. Rodríguez Vaquero, L. R. Tomassini Serravalle, M. C. Manca de Nadra and A.M. Strasser de Saad. *Food Control.* 21, 779 (2010).

[6] J. H. Chávez, P. C.Leal, R. A. Yunes, R. J. Nunes, C. R. M. Barardi, A. R. Pinto, et al. *Vet. Microbiol.* 116, 53 (2006).

[7] B. Suarez, A. L. Alvarez, Y. Dineiro Garcia, G. del Barrio, A. Picinelli Lobo and F. *Parra. Food Chem.* 120, 339 (2010).

[8] O. Escuredo, L. R. Silva, P. Valentão, M. C. Seijo and P. B. Andrade. *Food Chem.* 130, 671 (2012).

[9] P. M. Kris-Etherton, K. D. Hecker, A. Bonanome, S. M. Coval, A. E. Binkoski, K. F. Hilpert, et al. *Am. J. Med.* 113 (9B), 71S (2002).

[10] A. Shimizu-Ibuka, H. Udagawa, K. Kobayashi-Hattori, K. Mura, C. Tokue, T. Takita, et al. *Biosci. Biotechnol. Biochem.* 73 (1), 205 (2009).

[11] S. Pascual-Teresa, D. A. Moreno and C. García-Viguera. *Int. J. Mol. Sci.* 11, 1679 (2010).

[12] L. Hooper, P. A. Kroon, E. B. Rimm, J. S. Cohn, I. Harvey, K. A. Le Cornu, et al. *Am. J. Clin. Nutr.* 88, 38 (2008).

[13] E. E. Mulvihill, M. W. Huff. *Can. J. Cardiol.* 26 (A), 17 (2010).

[14] D. Ghosh and A. Scheepens. *Mol. Nutr. Food Res.* 53, 322 (2009).

[15] N. G. Amado, B. F. Fonseca, D. M. Cerqueira, V. M. Neto and J. G. Abreu. *Life Sci.* 86, 545 (2011).

[16] S. Prasad, K. Phromnoi, V. R. Yadav, M. M. Chaturvedi and B. B. Aggarwal. *Planta Med.* 76, 1044 (2010).

[17] J. Mojzis, L. Varinska, G. Mojzisova, I. Kostova and L. Mirossay. *Pharmacol. Res.* 57, 259 (2008).

[18] Y. Lin, R. Shi, X. Wang and Han-Ming Shen. *Curr. Cancer Drug Tar.* 8 (7), 634 (2008).

[19] R. M. Lopes, T. da Silveira Agostini-Costa, M. Aparecido Gimenes and D. Silveira. *J. Agr. Food Chem.* 59, 4321 (2011).

[20] M. L. Francisco and A. V. A. Resurrección. *Crit. Rev Food Sci.* 48, 715 (2008).

[21] Y. C. Chukwumah, L. T. Walker, M. Verghese, M. Bokanga, S. Ogutu, and K. Alphonse. *J. Agric. Food Chem.* 55, 285 (2007).

[22] J. Yu, M. Ahmedna, I. Goktepe and J. J. Dia. *J. of Food Compos. Anal.* 19, 364 (2006).

[23] P. M. Kris-Etherton, G. Zhao, A. E. Binkoski, S. M. Covaland, T. D. Etherton. *Nutr. Rev.* 59 (4), 103 (2001).

[24] M. Kris-Etherton, F. B. Hu, E. Ros and J. Sabaté. *J. Nutr.* 138, 1746S (2008).

[25] B. D. Craft, A. Kosińska, R. Amarowicz and R. B. Pegg. *Plant Foods Hum. Nutr.* 65, 311 (2010).

[26] C. E. Duncan, D. W. Gorbet and S. T. Talcott. *Food Res. Int.* 39, 898 (2006).

[27] J-C. Chang, Y-H. Lai, B. Djoko, P-L. Wu, C-D. Liu, Y-W Liu, et al. *J. Agric. Food Chem.* 54, 10281 (2006).

[28] P. Ribéreau-Gayon. *Les Composées phenoliques des Vegetaux.* Ed. Dunod, Paris, France (1968).

[29] I. Blázquez-Rojas. *Evaluación del potencial fenólico de hollejos de uva para la obtención de extractos colorantes y antioxidantes alimenticios y taninos enológicos.* Universidad de Castilla-La Mancha, Spain (2005).

[30] J-J. Macheix, A. Fleuriet and J. Billot. *Fruit phenolics.* CRC Press, Boca Raton, FL, USA (1990).

[31] M. C. Llaudy Fernández. *Contribución al estudio de los factores que afectan a la astringencia del vino tinto.* Universitat Rovira i Virgili, Spain (2006).

[32] D. M. Pereira, P. Valentão, J. A. Pereira and P. B. Andrade. *Molecules.* 14, 2202 (2009).

[33] J. Petersson and J. Dwyer. *J. Am. Diet. Assoc.* 98, 677 (1998).
[34] P. M. Aron and J. A. Kennedy. *Mol. Nutr. Food Res.* 52, 79 (2008).
[35] S. C. de Pascual-Teresa Fernández. *Análisis de taninos condensados.* Universidad de Salamanca, Spain (1999).
[36] J. Pokorny, N. Yanishlieva and Michael Gordon. *Antioxidants in Food. Practical application.* Woodhead Publishing Ltd, Cambridge, England and CRC Press, Boca Raton, FL, USA (2001).
[37] M. L. Francisco and A. V. A. Resurrección. *J. Food Compos. Anal.* 22, 16 (2009).
[38] R-S. Chen, P-L. Wu and R. Y-Y. Chiou. *J. Agric. Food Chem.* 50, 1665 (2002).
[39] D. J. Daigle, E. J. Conkerton, T. H. Sanders and, A. C. J. Mixon. *Agric. Food Chem.* 36, 1179 (1988).
[40] G. C. Yen, P. D. Duh and C. L. J. Tsai. *J. Agric. Food Chem.* 41, 67 (1993).
[41] V. S. Sobolev and R. J. Cole. *J. Agric. Food Chem.* 47, 1435 (1999).
[42] M. Monagas, I. Garrido, R. Lebrón-Aguilar, M. C. Gómez-Cordovés, A. Rybarczyk, R. Amarowicz et al. *J. Agric. Food Chem.* 57, 10590 (2009).
[43] J. Yu, M. Ahmedna and I. Goktepe. *Food Chem.* 90, 199 (2005).
[44] Y. Chukwumah, L. Walker, B. Vogler and M. Verghese. *J. Agric. Food Chem.* 55, 9266 (2007).
[45] S. T. Talcott, C. E. Duncan, D. D. Pozo-Insfran and D. W. Gorbet. *Food Chem.* 89, 77 (2005).
[46] V. S. Sobolev. *J. Agric. Food Chem.* 49, 3725 (2001).
[47] K. J. Dabrowski and F. W. Sosulski. *J. Agric. Food Chem.* 32,128 (1984).
[48] C. Edwards and R. N. Strange. *J. Chromatogr.* A. 547,185 (1991).
[49] C. Edwards, R. N. Strange and D. L. Cole. *Plant Pathol.* 44, 573 (1995).
[50] J. Liu, G. Wang and G. Wang. *Chin. Tradit. Patent Med.* 31, 1902 (2009).
[51] W. J. Yen, L. W. Chang and P. D. Duh. *Lebensm.-Wiss. u.-Technol.* 38, 193 (2005).
[52] H. Lou, H. Yuan, Y. Yamazaki, T. Sasaki, and S. Oka. *Planta Med.* 67, 345 (2001).
[53] S. V. Verstraeten, J. F. Hammerstone, C. L. Keen, C. G. Fraga and P. I. Oteiza. *J. Agric. Food Chem.* 53, 5041 (2005)
[54] M. M. Appeldoorn, J-P. Vincken, M. Sanders, P. C. H. Hollman and H. Gruppen. J. Agric. *Food Chem.* 57, 6007 (2009a).

[55] M. M. Appeldoorn, M. Sanders, J-P. Vincken, V. Cheynier, C. Le Guernevé, P. C. H. Hollman, et al. *Food Chem.* 117, 713 (2009b).
[56] H. Lou, H. Yuan, B. Ma, D. Ren, M. Ji and S. Oka. *Phytochem.* 65, 2391 (2004).
[57] S. A. Lazarus, G. E. Adamson, J. F. Hammerstone and H. H. Schmitz. *J. Agric. Food Chem.* 47, 3693 (1999).
[58] C. Manach, G. Williamson, C. Morand, A. Scalbert and C. Rémésy. *Am. J. Clin. Nutr.* 81, 230S (2005).
[59] G. Williamson and C. Manach. *Am. J. Clin. Nutr.* 81, 243S (2005).
[60] S. Déprez, I. Mila, J-F. Huneau, D. Tomé and A. Scalbert. *Antiox. Redox Signal.* 3, 957 (2001).
[61] M. Ibern-Gómez, S. Roig-Pérez, R. M. Lamuela-Raventós and M. C. de la Torre-Boronat. *J. Agric. Food Chem.* 48, 6352 (2000).
[62] J. M. van Amelsvoort, K. H. Van Hof, J. N. Mathot, T. P. Mulder, A. Wiersma and L. B. Tijburg. *Xenobiotica,* 31, 891 (2001).
[63] A. Moure, J. M. Cruz, D. Franco, J. M. Dominguez, J. Sineiro, H. Dominguez, et al. *Food Chem.* 72, 145 (2001).
[64] H. Lou, H. Yuan, B. Ma, D. Ren, M. Ji, S. Oka. *Phytochemistry.* 65, 2391 (2004).
[65] S. V. Verstraeten, C. L. Keen, H. H. Schmitz, C. G. Fraga and P. I. Oteiza. *Free Radical Bio. Med.* 34 (1), 84 (2003).
[66] V. Nepote, M. G. Mestrallet and N. R. Grosso. *J. Food Sci.* 69 (7), 295 (2004).
[67] M. Dueñas, S. González-Manzano, A. González-Paramás, C. Santos-Buelga. J. Pharmaceut. *Biomed.* 51, 443 (2010).
[68] J. Wang, X. Yuan, Z. Jin, Y. Tian and H. Song. *Food Chem.* 104, 242 (2007).
[69] T. S. Ballard, P. Mallikarjunan, K. Zhou and S. F. O'Keefe. *J. Agric. Food Chem.* 57, 3064 (2009).
[70] X. Wu, G. R. Beecher, J. M. Holden, D. B. Haytowitz, S. E. Gebhardt and R. L. Prior. *J. Agric. Food Chem.* 52, 4026 (2004).
[71] K. Yanagimoto, K. Lee, H. Ochi and T. Shibamoto. *J. Agric. Food Chem.* 50, 5480 (2002).
[72] L. Kang, W. Heng, A. Yuan, L. Baolin and H. Fang. *Biochimie.* 92, 789 (2010).
[73] M. Yamakuchi, C. Bao, M. Ferlito and C. J. Lowenstein. *Biol. Chem.* 389, 935 (2008).
[74] S. de Pascual-Teresa, D. A. Moreno and C. García-Viguera. *Int. J. Mol. Sci.* 11, 1679 (2010).

[75] D. F. Fitzpatrick, R. C. Fleming, B. Bing, D. A. Maggi and R. M. O'Malley. *J. Agric. Food Chem.* 48, 6384 (2000).
[76] J. I. Ottaviani, L. Actis-Goretta, J. J. Villordo and C. G. Fraga. *Biochimie.* 88, 359 (2006).
[77] M. Das and D. K. Das. *Mol. Aspects Med.* 31, 503 (2010).
[78] M. T. Garcia-Conesa, S. Tribolo, S. Guyot, F. A. Tomas-Barberan and P. A. Kroon. *Mol. Nutr. Food Res.* 53, 266 (2009).
[79] S. Heptinstall, J. May, S. Fox, C. Kwik-Uribe and L. Zhao. *J. Cardio.Pharm.* 47, 197 (2006).
[80] M. Pinet, A. Castell, I. Baiges, G. Montagut, L. Arola and A. Ardévol. *Compr. Rev. Food Sci.* F. 7, 299 (2008).
[81] S. Uchiyama, Y. Taniguchi, A. Saka, A. Yoshida and H. Yajima. *Nutrition.* 27, 287 (2011).
[82] H.-S. Moon, H.-G. Lee, Y.-J. Choi, T.-G. Kim and C.-S. Cho. *Chemico-Biological Interactions.* 167, 85 (2007).
[83] D. A. Moreno, N. Ilic, A. Poulev and I. Raskin. *Life Sci.* 78, 2797 (2006).
[84] J. M. Pavese, R. L. Farmer and R. C. Bergan. *Cancer Metast. Rev.* 29, 465 (2010).
[85] Z. P. Chen, J. B. Schell, Chi-Tang Ho and K. Y. Chen. *Cancer Letters.* 129, 173 (1998).
[86] T. Nakazato, K. Ito,Y. Ikeda and M. Kizaki. *Clin. Cancer Res.* 11, 6040 (2005).
[87] M. Kurzawa-Zegota, M. Najafzadeh, A. Baumgartner and D. Anderson. *Food Chem. Toxicol.* 50, 124 (2012).
[88] G.-C. Yen and P.-D. Dhu. *Biosci. Biotech. Biochem.* 60 (10), 1698 (1996).
[89] M. Athar, J. H. Back, X. Tang, K. H. Kim, L. Kopelovich, D. R. Bickers et al. *Toxicol. Appl. Pharmacol.* 224 (3), 274 (2007).
[90] M. Singh, M. Arseneault, T. Sanderson, V. Murthy and C. Ramassamy. *J. Agric. Food Chem.* 56, 4855 (2008).
[91] D. Brambilla, C. Mancuso, M. R. Scuderi, P. Bosco, G. Cantarella, L. Lempereur, et al. *Nutr. J.* 7, 1 (2008).
[92] S. Mandel and M. B. H. Youdim. *Free Radical Bio.Med.* 37 (3), 304 (2004).
[93] L. Rossi, S. Mazzitelli, M. Arciello, C. R. Capo and G. Rotilio. *Neurochem. Res.* 33, 2390 (2008).
[94] L. Letenneur, C. Proust-Lima, A. Le Gouge, J.Dartigues and P. Barberger-Gateau. *Am. J. Epidemiol.* 165 (12), 1364 (2007).

[95] T. Pan, J. Jankovic and W. Le. *Drugs Aging.* 20 (10), 711 (2003).
[96] C. Mancuso, T. E. Bates, D. A. Butterfield, S. Calafato, C. Cornelius, A. De Lorenzo, et al. *Expert Opin. Investig. Drugs.* 16 (12), 1921 (2007).
[97] O. Weinreb, S. Mandel, T. Amit and M. B. H. Youdim. *J. of Nutr. Biochem.* 15, 506 (2004).
[98] Y. Levites, T. Amit, S. Mandel and M. B. H. Youdim. *FASEB J.* 17, 952 (2003).
[99] T. S. Anekonda. *Brain Res. Rev.* 52, 316 (2006).
[100] V. Vingtdeux, U. Dreses-Werringloer, H. Zhao, P. Davies and P. Marambaud. *BMC Neurosci.* 9 (2), S6 (2008).

# PART 2. PEANUT ALLERGY

In: Peanuts: Bioactivities and Allergies
Editors: I.L. Expósito, A.B. Blázquez
ISBN: 978-1-62417-387-5


*Chapter 5*

# LEGUMES ALLERGY: CLINICAL SYMPTOMS AND CROSS-REACTIVITY

***M. Reche*****[*]**, ***T. Valbuena and C. Y. Pascual***
Allergy Service, Infanta Sofía Universitary Hospital,
San Sebastián de los Reyes, Madrid, Spain

## ABSTRACT

Legumes are a relatively cheap and excellent source of high-biological-value proteins and other nutrients and also they are an important source of protein for a great part of the world's population, especially in the developing countries. They are widely used around the world, constitute an important part of healthy vegan's diets, and are common ingredients in many processed food products. Legumes are one of the most frequent plant foods associated with allergic reactions in children. Among food sources, belonging to the *leguminoseae* family, peanut and soybean are the two major ones involved in immediate hypersensitive responses in countries such as the United States, Japan and the United Kingdom. However, other legumes, such as lentils and chickpeas, seem to be an important cause of IgE-mediated allergy in the Mediterranean area. Most of the allergens already characterized from legumes are related with the 2S storage proteins superfamily. Fractionation of legume proteins by solubility procedures shows three

---

[*] E-mail address: marta.reche@salud.madrid.org.

fractions: globulins (predominant), albumins and glutelins. The high content of globulins in legume seeds evidence that the biological function is mostly as storage proteins. The globulin fraction can be separated by ultra centrifugation in vicilin and legumin. Clinical relevance of cross-reactivity between legumes is well known. The introduction of molecular biology techniques to the study of allergens has prompted the analysis of their primary structure, biological function, homology, and thus have provided a wider future to the study of cross-reactivity among allergens. The ImmunoCAP Molecular Allergology technique allows knowing the clinical relevance of the different sensitizations of the legume´s allergens.

## INTRODUCTION

Legumes are dicotyledonous plants belonging to the *Fabaes* order. The main distinctive characteristic is their fruit (legumen, seeds contained in pods). This botanical order is formed by three families: *Mimosaceae, Caesalpiniaceae* and *Papilionaceae* or *Fabaceae.* The *Papilionaceae* family includes the most important allergenic species: *Lens culinaris* (lentil), *Cicer arietinium* (chickpea), *Pisum sativum* (pea), *Arachis hipogea* (peanut), *Phaseola vulgaris* (bean) *Lupinus angustifolius* (lupine) and *Glycine max* (soy) (Table 1) [1, 2].

Legumes are a relatively cheap and excellent source of high-biological-value proteins and other nutrients in addition to an important source of protein for a great part of the world's population, especially in the developing countries.

They are widely used around the world, constitute an important part of healthy vegans' diets, and are common ingredients in many processed food products [3]. The leader in world production of seed legumes other than soybeans, is Asia, being India the highest collector of lentil, chickpeas and peanut and the U. S. leader in soybeans production. Countries like India use to eat legumes as the first food after breast feeding.

Legumes are one of the most frequent plant foods associated with allergic reactions in children.

Among food sources, belonging to the *leguminoseae* family, peanut and soybean are the two major ones involved in immediate hypersensitive responses in countries such as the United States, Japan and the United Kingdom [4, 5]. However, other legumes, such as lentils and chickpeas, seem to be an important cause of IgE-mediated allergy in the Mediterranean area [2, 6, 7].

Previous reports have shown that in Spain, lentils are the legumes most commonly implicated in food allergy [8] although in the Mediterranean area other members of the *leguminosae* family such as chickpeas, green peas, various types of beans and in lesser amount, peanuts and soy are also consumed.

In recent years, it has been suggested that lupins should be regarded as hidden allergens as their flour is commonly used in many European countries as an "improving" ingredient in the production of bread, biscuits, crusts, pasta, and other bakery products and/or as soya substitutes [9].

**Table 1. Principal components of the *Papilionaceae* family**

| Common name | Gender and Species |
|---|---|
| Lentil | *Lens sculenta or culinaris* |
| Chick pea | *Cicer aretinum* |
| Peanut | *Arachis hypogea* |
| Bean | *Vicia faba* |
| Pea | *Pisum sativum* |
| Soy | *Glycine max* |
| Lupin | *Lupinus angustifolius* |
| Green bean | *Phaseolus vulgaris* |

## Clinical Symptoms of Legumes Allergy

Allergy to legumes may be mediated by primary sensitization via ingestion, or inhalation, or may be acquired after primary sensitization to pollen or to other foods. Allergic reactions to food may affect many different organs and systems, being the gastrointestinal tract, including the oral mucosa and skin commonly involved. [6] Airways are less commonly affected and the cardiovascular system may be affected in systemic anaphylactic reactions.

The clinical symptoms which can be found in a food-allergic patient are:

- *Oral allergy symptoms*: pruritus of lips, tongue and oropharyngeal mucosa, swelling of the lips and perioral rash may be the initial manifestation of a systemic reaction to food, but also can be the only manifestation after the legume ingestion for a sensitized patient.
- *Gastrointestinal symptoms*: are common in legumes allergy and consist of nausea, vomiting, abdominal pain and diarrhoea.

- *Respiratory symptoms*: bronchospam and stridor are common in acute allergic reaction. Chronic asthma and rhinoconjunctivitis may rarely be caused by legumes allergens, but legumes have been involved in others respiratory syndromes. In 1990 one legume, soybean, was listed as an agent causing occupational disease when a soybean-dependent asthma epidemic occurred in Barcelona and Cartagena (Spain) after a dust discharge into the air. Years later, new antigens of soybean were described as cause of hypersensitivity pneumonitis (HP) in a research agronomist who was exposed to them for a long time. Other legumes have been involved in occupational asthma.
- *Skin manifestations (acute urticaria/angioedema)*: development of an urticarial rash and erythematus swelling of all the face and lips is a common feature of IgE- mediated allergy to legumes. This usually occurs within a few minutes to 2 hours of ingestion and the cause - effect relationship is not difficult to establish.
- *Anaphylactic reactions:* are caused by IgE-mediated hypersensitivity reactions. These involve the classic pathway of mast cell and basophil degranulation with the release of mediators, as histamine. Any food can cause anaphylaxis but peanut is the most common cause of food-induced anaphylaxis and death. In same cases, exercise following food consumption increases the risk of anaphylaxis. An anaphylactic reaction usually occurs within minutes of exposure to legumes, but may be delayed by up to 2 hours. If the patient is very sensitive, a rash and itching may develop after contact with the skin or oral mucosa. Usual symptoms are acute urticarial rash, swelling of lips and tongue, followed by gastrointestinal symptoms of colic, vomiting and diarrhoea and respiratory symptoms of throat tightness, wheezing and difficult in breathing. Hypotension, collapse and death may occur.

Peanut is the most common food involved in fatal allergic reactions in western countries. Accidental peanut consumption may cause severe allergic reactions in susceptible individuals. In contrast to other common food allergies in childhood, peanut allergy is often a lifelong disease. The symptoms upon peanut ingestion might vary from mild reactions, such as urticaria or oral allergy syndrome, to respiratory distress and severe systemic reactions needing medical care, such as anaphylactic shock. [10]

**Table 2. Legume allergens**

| Allergenic Protein | Plant family | Superfamily | MW (KDa) |
|---|---|---|---|
| **Peanut** *(Arachis hypogea)* | | | |
| Ara h 1 | 7/8S globulins (vicilins) | Cupin | 64 |
| Ara h 2 | 2S albumins (conglutin) | Prolamin | 17 |
| Ara h 3 | 11S globulins (legumins) | Cupin | 60 |
| Ara h 4 | 11S globulins (legumins) | Cupin | 37 |
| Ara h 5 | Profilin (Bet v 2-like) | Profilin | 15 |
| Ara h 6 | 2S albumins (conglutin) | Prolamin | 15 |
| Ara h 7 | 2S albumins (conglutins) | Prolamin | 15 |
| Ara h 8 | Bet v 1 family (PR-10, Bet v 1-like) | PR proteins | 17 |
| Ara h 9 | nsLTPs | Prolamin | 9.8 |
| Ara h 10 | Oleosin plant lipid storage bodies | Oleosin | 16 |
| Ara h 11 | Oleosin plant lipid storage bodies | Oleosin | 14 |
| **Soybean** *(Glycine max)* | | | |
| Gly m 1 | nsLTPs | Prolamin | 7 |
| Gly m 2 | Defensin | | 8 |
| Gly m 3 | Profilin | Profilin | 14 |
| Gly m 4 | Bet v 1 family (PR-10, Bet v 1-like) | PR proteins | 17 |
| Gly m 5 | 7S globulin, beta-conglycinin (vicilin) | Cupin | |
| Gly m 6 | 11S globulin, glycinin (legumin) | Cupin | |
| **Lentil** *(Lens culinaris)* | | | |
| Len c 1 | Gamma-vicilin subunit | Cupin | 47 |
| Len c 2 | Seed-specific biotinylated protein | | 66 |
| Len c 3 | nsLTP | Prolamin | 9 |
| **Pea** *(Pisum sativum)* | | | |
| Pis s 1 | Vicilin | Cupin | 44 |
| Pis s 2 | Convicilin | Cupin | 63 |
| **Lupin** *(Lupinus angustifolius)* | | | |
| Lup an 1 | Beta-conglutin (vicilin) | Cupin | 55-61 |
| **Green bean** *(Phaseolus vulgaris)* | | | |
| Pha v | nsLTP | Prolamin | 8.8-9 |

MW: molecular weight. PR: proteins related. Ns LTP: nonspecific lipid transfer protein.

(Modified from Nicolau N, Custovic A. Molecular diagnosis of peanut and legume allergy. Curr Opin Allergy Clin Immunol 2011).

## CROSS-REACTIVITY OF LEGUMES ALLERGY

The clinical cross-reactivity between legumes exists. [8] Several studies have shown cross-reactivity between peanut and lupine. Peeters et al [11] showed that 71% (5/7) of peanut-allergic patients had a positive skin prick test (SPT) to lupine. They studied the patterns and frequencies of sensitization to lupine, pea and soy in a peanut-allergic population. Thirty-nine unselected peanut-sensitized patients were evaluated by skin prick tests and ImmunoCAP to lupine, pea and soy. Clinical reactivity was measured by double-blind placebo-controlled food challenges (DBPCFC). Eighty-two percent of study population was sensitized to lupine, 55% to pea, and 87% to soy, but clinical relevant sensitization to lupine, pea and soy occurred in 35%, 29% and 33% respectively of the study population. Moneret Vautrin *et al*, [12] showed that out of 24 peanut-allergic patients six revealed a positive SPT response to both peanut and lupine, and five out of six patients tested, reacted to lupine as determined by DBPCFC. This fact suggests that the use of lupine is a risk especially for peanut-allergic patients. Fiocchi *et al* [13] included 12 selected peanut allergic patients in a trial where lupine-enriched pasta and placebo pasta were administered in a DBPCFC protocol. Positive clinical reactions were observed in two children, the eliciting dose being 0.2 and 6.4 g of pasta, corresponding to 50 mg and 1.6 g of lupine proteins, respectively. β-conglutin was the protein most involved in SPT positive. Authors concluded that Lupine-enriched pasta can be tolerated by most subjects suffering from peanut allergy, but a sizeable minority (2/12 of them in this case) can develop potentially dangerous clinical reactions. Severe allergic reactions to lupine have increased in recent years with the growing use of lupine flour as wheat or soy substitute in manufactured foods. Since lupine often appeared as an unlabeled ingredient, the substance was identified as a hidden food allergen. The first case of lupine allergy was reported by Hefle *et al* [14] in 1994 in a 5-year-old girl with a peanut allergy who developed urticaria and angioedema after eating spaghetti fortified with lupine flour. Since then, cases of allergy to lupine flour as a hidden allergen in a range of foods such as ready-made tomato sauce, biscuits, onion rings, pizza and chicken bouillon cubes have been reported by Las Marinas *et al*, [15] and Rojas Hijazo *et al.* [16] Although the majority of case reports were produced by food ingestion, Moreno-Ancillo *et al* [17] described respiratory symptoms in a child after the inhalation of lupine dust. Prieto A *et al*, [18] report a case of recurrent anaphylaxis caused by lupine as a hidden allergen proceeded by former sensitization via inhalation. The authors described a 54-year-old woman that presented generalized urticaria, lip angioedema, itching in the ears and throat, dysphonia, and cough immediately after eating lupine seeds. She avoided lupine after this episode and tolerated other legumes such as peanut, lentils, chickpeas, green beans, peas and soy. Besides, when the patient visited her parent's home, at any time of the year, she developed eyes and nose itching sneezing, watery rhinorrhea, nasal stuffiness and cough at night. Her mother had adopted the habit of grinding lupine seeds to use as sweetener. In this patient, skin prick test with commercial extracts of peanut, lentil, chickpea, green bean, white been, pea, soy, tomato, apple, mustard, wheat, barley and rye flour, almond, hazelnut, walnut, mil, egg, latex and Anisakis simple were all negative. Prick-to-prick tests were positive for lupine seed, and ready-made tomato sauce, and negative for spices (pepper, oregano, basil and nutmeg). Specific-IgE was detected for lupine 2.62 Ku/l (CAP system Pharmacia, Upssala, Swedden), The manufacturers of ready-made tomato sauce confirmed the presence of lupine flour in this product. An inhalation exposure test with ground lupine seeds was performed and was considered positive. With the coming into force of the European labelling Directive 2006/142/EC on December 23, 2008, lupine is now considered a potentially allergenic ingredient that must be specifically included by food manufacturers on their labels. Although Lupine allergy is generally associated with sensitization to other legumes, mainly peanut, there have been reports of clinical symptoms due to lupine between 4% and 28% of patients with peanut allergy, with rates varying according to the gastronomic habits and date of authorization for the use of lupine flour in manufactured foods in each country.

In Spain, legume allergy, mainly to lentils and chickpeas, is the fifth most common cause of food allergy in children. [8] Clinical relevance of cross-reactivity between legumes is well known since the publication of Bernhisel-Broadbent [19] and Pascual *et al.* [20] Martinez San Irineo *et al* [21] studied fifty-four children with clinical allergy to legumes. Cross-reactivity was evaluated by ELISA inhibitions experiments and oral food challenges with legumes. The oral legumes challenges demonstrated that 37 children (69%) were allergic to 2 or more legumes (median 3 legumes). The most frequent associations were allergy to lentils and chickpeas (57%), allergy to lentils and peas (54%), and allergy to lentils, chickpeas and peas (43%). Symptomatic hypersensitivity to chickpeas is frequently associated with lentil allergy and we have not found patients monosensitized to chickpeas as we did with lentils. Cross-reactivity between legumes other than lentils and chickpeas are widely known by *in vitro* studies [22, 23].

# ALLERGENS AND ALLERGENICITY

The 19.6% of soy and the 47.9% of peanuts composition is lipids, but only 5% in chickpeas and 0,8% in lentils are so. The protein content in these 4 legumes are 39% in soy, 24% in peanuts, 26% in chickpeas, and 28% in lentil [24]. Legume proteins are deficient in S-containing aminoacids (cystine, methionine and tryptophan). Some of them contain toxic substances such as cyanogenic glycosides (lime beans) and anti-nutritive factors as proteinase inhibitors and lectins, but the food processing by heating inactivates them. [25]

In general, food allergens are low-molecular-weight proteins or glycoproteins (< 70kD) that are soluble in aqueous solution and abundant in food. These molecules are generally resistant to proteases, heat, and denaturants, thus preventing destruction during food preparation and digestion. [26] Many allergenic proteins have been recently characterized and used within the context of molecular diagnosis or component-resolved diagnosis to facilitate more accurate diagnosis of food allergy. Most of the allergens already characterized from legumes are related with the superfamily of 2S storage proteins (Table 2). Fractionation of legume proteins by solubility procedures shows three fractions: globulins (predominant), albumins and glutelins. The high content of globulins in legume seeds evidence that the biological function is mostly as storage proteins. The globulin fraction can be separated by ultra centrifugation in vicilin and legumin. Vicilins from this group of proteins is an important common allergen in legumes clinical allergy and maybe it could be used as indicator or marker in diagnosis of this type of sensitization. The knowledge of new allergens in legumes and the assessment of their high or low allergenicity can be of great interest for a future manipulation of legume allergens by protein engineering in order to produce less allergenic seeds.

## Peanut Allergens

Peanut (*Arachis hypogea*) is an annual plant that belongs to the *leguminosae* (legumes) family and is native to South America. Multiple varieties, including Virginia, Spanish, and Runners are grown in the USA, with more than 40% of the American peanut crop consumed as peanut butter.The allergenic portion of peanut is similar among the different varieties of peanuts and has been identified as the protein of the cotyledon. [27, 28] The major peanut allergens are homologous to the seed storage proteins of the conglutin, vicilin, and glycinin families [29]. Eleven peanut allergens have been recognized by the International Union of Immunological Societies (IUIS) Nomenclature Subcommittee [30]. Although peanut allergens will be extensively revised in chapter 6, a brief description of their principal characteristics will be given on this paragraph.

In the early 1990s, Ara h 1 and Ara h 2 were identified and are the most extensively studied [31, 32]. Ara h 1-3 have been considered the major peanut allergens, [33] although recent reports suggest an important role for Ara h 6. [34]

Ara h 1 is a member of the 7/8S globulins (vicilins) of seed storage proteins that belongs to the cupin superfamily. The gene for Ara h 1 was cloned and sequenced, revealing significant sequence homology with plant seed storage proteins in the vicilin family [31]. Ara h 2, 6 and 7 are members of the 2S albumins (conglutin) of the seed storage proteins that belong to the prolamin superfamily. Cloning and sequencing the Ara h 2 gene revealed sequence homology with the conglutin family of seed storage proteins. Ara h 2 has also a high sequence homology with Ara h 6. As a result of their stability to heat and gastrointestinal digestion, many allergens of the prolamin superfamily are important and may account for severe allergic reactions [32]. Ara h 3 and 4 are nearly identical isoforms of the 11S globulins (legumins/glycinins) of seed storage proteins that belong to the cupin superfamily. [35] The aminoacid sequence of Ara h 3 shows significant homology to the glycinin family of seed storage proteins [36]. Ara h 5 belongs to the profilin family. Profilins show high sequence homologies even from distantly related plants and are also known as pan-allergens involved in crossreactions between pollen and plant foods [37]. Ara h 8 belongs to the pathogenesis-related protein family (PR)-10 and it is of relevance to peanut-allergic patients with birch pollen allergy because of the crossreactivity to homologous Bet v 1 allergen [38, 39]. Ara h 9 belongs to the non specific lipid transfer proteins (nLTPs), from the prolamin superfamily [40, 41]. Ara h 10 and 11 or oleosins have recently been recognized and their allergenic properties are largely unknown. Allergenic oleosins have been found in legumes, nuts and seeds.

A number of factors related to the harvesting and processing of peanuts may affect the allergenic properties of these proteins. For example, roasting enhances the allergenicity of Ara h 1-3, whereas boiling or frying appears to reduce their allergenic properties. Ara h 5 and 8 are heat labile and cook them has an adverse effect on their allergenicity. [42] These food processing methods vary by geographic region. For instance, despite a high rate of peanut consumption, the prevalence of peanut allergy in China is much lower than in USA. In China, peanuts are usually fried or boiled, whereas in the USA peanuts are frequent dry roasted.

Using recombinant peptides from Ara h 1, 2 and 3, Beyer *et al.* [43] have demonstrated differences in epitope recognition between patients with symptomatic allergy compared with those who were sensitized, but tolerant. Patients with IgE binding to a greater number of epitopes had a history of more severe reactions. Similarly, Chinag *et al.* [44] evaluated specific IgE responses to the major peanut allergens Ara h 1-3 in 31 peanut-sensitized Asian children, 27 of the children were considered peanut allergic. They found that reactivity to *Ara h 1* and Ara h 2 was predictive of peanut allergy. Nicolau *et al.* [45] studied the allergy or tolerance in children sensitised to peanut using the measurement of IgE response to different peanut components. They found that peanut-allergic patients tended to have a higher IgE response against the major peanut components, Ara h 1-3. Anyhow, Ara h 2 offered the best discrimination between peanut-allergic and peanut-tolerant patients.

Vereda *et al.* [46] demonstrated markedly different clinical and immunological patterns of peanut allergy in patients from three different geographical areas. Peanut allergic patients (challenge or convincing history) were recruited in the USA (n=30), Spain (n=50) and Sweden (n=35). American patients frequently had IgE antibodies to r*Ara h 1* to 3 (56.7% to 90%) and often presented severe symptoms. Spanish patients recognized these 3 recombinant peanut allergens less frequently (16% to 42%), and they were more often sensitized to the lipid transfer protein rAra h 9 (60%), and typically had peanut allergy after becoming allergic to other plant-derived foods. Swedish patients detected r*Ara h 1* to 3 more frequently than Spanish patients (37.1% to 74.3%) and had the highest sensitization rate to the Bet v 1, homologue to rAra h 8 (65.7%) as well as to rBet v 1 (82.9%). American patients became allergic around 1 year of age, whereas Spanish and Swedish patients developed peanut allergy at the age of 2 years or later. These results have demonstrated heterogeneity in the clinical and immunological phenotype of peanut allergy in distinct geographical areas, which might reflect exposures to different environmental pollen and dietary traditions and emphasize the importance of the molecular diagnosis.

Asarnoj *et al.* [47] investigated, using microarrays, the IgE reactivity to peanut allergen components in school-age children in relation to pollen sensitization and reported peanut allergy. They found that children with concomitant sensitization to peanut and birch reported fewer symptoms to peanuts than children with sensitization to peanut only; peanut allergy was more common and severe in those with reactivity to Ara h 1-3 than those sensitised to Ara h 8 or to pollen.

## Soybean Allergens

Soybean *(Glycine max)* is a member of the legume family with a particularly rich content in proteins of high-biological value and therefore an important nutrient. Allergenicity of soybean proteins has been documented since the 1980s. Soybean exposure is widespread in Asia and the USA but its consumption has increased in the rest of the world. Concretely, in Europe in vegetarian cuisine and because of food technological benefits achieved by adding soy proteins to various processed food products. Soy is consumed as soy oil, soy flour, soy milk, soy drinks, soy flakes or as fermented soybean products such as Miso, Okara, soy sauce (Tamari, Shoyu), Tempeh or tofu. As soy is a good and cheap protein source, it may be part of a wide variety of processed foods such as meat products, sausages, bakery goods, chocolate or breakfast cereals [48] (Table 2). Soybeans contain about 40% of protein, the majority of which is composed of the 2 storage proteins β-conglycinin and glycinin. [49] Four other proteins are officially accepted as allergens. At least an additional 12 proteins from soybeans have been reported as IgE-reactive proteins [48]. Gly m 1 and Gly m 2 are components of the soybean hull and have been associated with respiratory symptoms through inhalation of soy particles [48, 50]. Gly m 3 is a profilin and is considered a minor allergen [51]. Gly m 4 is a Bet v 1 homologue, belongs to the PR-10 family and is a common allergen in patients with birch pollen and soy allergy. [52, 53]. Gly m 5 (β-conglycinin) and Gly m 6 (glycinin) are 7S and 11S globulins (cupin superfamily) and are a major soybean storage proteins. Sensitization to theses two allergens has recently been suggested as a potential marker of severe allergic reactions among patients with soy allergy. [52-54].

## Lupine Allergens

Lupine is considered to be a rich source of protein with notable content of lysine and has been increasingly used in bakery, confectionary, snacks and pastry products due to its multifunctional properties. In addition it has potential hypocholesterolemic and hypoglycemic properties [9]. The conglutins α (legumin like) and β (vicilin like) are the most abundant allergenic proteins in *Lupinus* species. *Lup an 1*, a β-conglutin is the major allergen identified in *L. angusitfolius* and shares sequence homology with Ara h 1. Guillamon *et al.* [55] have described Lup-1 (β-conglutin), (34.5 kDa) and Lup -2 (α-conglutin fraction), (20 kDa) as major allergens in *L. albus* which show high-amino-acid sequence homology with the major allergens from peanut, pea, lentil and soybean. Dooper *et al*, described IgE cross-reactivity between the 2S albumins Ara h2 and δ-conglutin, and Ara h1 and β-conglutin, which are possibly based on homologies between phylogenetically related proteins. Ara h 2 was the most potent inhibitor of IgE binding of lupine conglutins.

## Green Bean Allergens

Green bean is a leguminous plant belonging to *Fabaceae* family, genus *Phaseolus,* species *vulgaris.* This fruit is constituted by a slender green pod with small seeds inside, different from other bean species, the entire pod is edible. Zoccatelli *et al.* [56] have recently characterized a non-specific LTP, identified as a major allergen of green bean, *Pha v 3*. In this study, a 10 kDa protein in green bean extract was recognized by 80% of the sera from 10 Spanish patients reporting adverse reactions to green bean, confirmed by oral challenge.

## Pea Allergens

Sánchez-Monge *et al.* [22], studied serum pool or individual sera from 18 patients with pea allergy to detect binding proteins from pea seeds by immunoblot inhibition assays. They found that the IgE immunodetection of crude pea extracts revealed that convicilin (63kDa), as well as vicilin (44kDa) and one of its proteolytic fragments (32kDa), reacted with more than 50% of the individual tested. They conclude that the vicilin and convicilin are potential major allergens from pea seeds.

**Table 3. Algorithm diagnosis for soy allergy**

| f14 | Gly m 4+ | Gly m5/ Gly m6 | Symptoms |
|---|---|---|---|
| + | - | + | Risk of Severe reactions |
| +/- | + | - | Mild reactions |

## Lentil Allergens

Lentils (*Lens culinaris* or *sculenta*), members of the *Leguminosae* family, were cultivated by Persians and Egyptians 2500 years BC. The consumption as common food in the rest of the Mediterranean area was due to Roman Empire facilities. Lentils are an annual plant with flatterened edible seeds, used in soups, mixed with cereal such as rice, or grounded as flour. Lentils are consumed after a heat treatment, usually boiling. Heating can produce inactivation, modification, or both of the putative allergens. That is why the two different types of allergens Len c 1 and Len c 2 were identified from boiled lentils. [57, 58] *Len c 1* is a protein of approximately 47 kDa identified as a vicilin chain. Three genetic isoforms of this allergen were described: Len c 1.0101, Len c 1.0102 and Len c 1.0103. The lentil allergen Len c 1 has 45% of identical residues with the corresponding part of the sequence in *Ara h 1* from peanut. A structural relationship between Len c 1 and several allergens of the vicilin family, including Jug r 2, Ana o 1, Ses i 3, Ara h 1, and soybean conglycinin were described. More than 70% of the patients with lentil allergy recognized the purified Len c 1. The location of the IgE-binding sequential epitopes of the major lentil protein Len c 1 has been recently reported [23]. *Len c 2* is a 66 kDa protein corresponding to a seed specific biotinylated protein. *Len c 3* has been recently characterized as a nsLTP. [59]

From the CAP inhibition of lentils with chickpeas, soybeans and peanuts it has been observed cross-inhibition but it is only partial, and it has been confirm by blot-inhibition. By immunoblotting it has been demonstrated that lentil and chickpeas inhibit each other more than they do to peanut and soybeans. The same can be observed between soybeans and peanut, related to lentils and chickpeas [8]. In our experience, Len c 1 can be a good model regarding legumes allergy, especially in those areas with a considerable intake of lentils and very low consume of peanut in children. A CAP-inhibition with Lenc1 and Len c 2 in fluid phase shows the highest inhibition for lentil, but Len c 1 inhibited more than 50% to peanut at only 100 ng/ml. Len c 2 inhibited lentils and peanut in a similar way, both lentil allergens inhibited soybean by CAP in a 30% of the cases at maximum dose of 100 ng.

## Chickpea Allergens

The chickpea is a legume consumed in many Mediterranean countries. Martinez *et al.* [21] studied the specific IgE to raw and boiled extracts, measured by ELISA, in 19 patients with clinical sensitivity to chickpea. The allergenic composition of both extracts was analyzed by SDS-PAGE and immunoblotting. There were no significant differences between specific IgE levels to raw and boiled extracts. Multiple IgE binding proteins/peptides were detected in both extracts in the molecular weight range of 10-106 kDa. Chickpea extracts contain a majority of heat-stable allergens, which could be responsible for the clinical sensitivity to chickpea.

## Molecular Allergology

The study of the relationship between allergens is of great interest for clinical and basic researchers in allergy. Cross-reactions between allergens have often been reported in recent years [60-64], but it is not a new finding in the history of allergy. Despite of large amount of publications on cross-reactivity, the complete cellular and molecular basis of these phenomena is still unknown. The introduction of molecular biology techniques to the study of allergens has prompted the analysis of their primary structure, biological function, homology, and thus have provided a wider future to the study of cross-reactivity among allergens. The ImmunoCAP Molecular Allergology technique allows knowing the clinical relevance of the different sensitizations to legumes' allergens. Table 3 and 4 show the algorithm disgnosis table for soy and peanut respectively. Regarding soy allergy, if the patient has positive IgE to f14 and Glym5 and Glym6, there is a high risk to severe reactions with soy ingestion or contact. However, if the determination od Glym5 and Glym6 are both negative, the risk of severe reactions is low (Table 3). In the case of Peanut Allergy, the absence of IgE binding to Ara h1, Ara h2 and Ara h3, nearly excluded the possibility of severe reactions with peanut (Table 4). Hence, the study of food allergy by components allows not only to diagnose food allergy but to predict the severity of a patient´s specific food sensitization. The study by components opens up a new therapeutic way to design specific immunotherapies according to the patient´s allergic profile.

**Table 4. Algorithm diagnosis for peanut allergy**

| f13 | Ara h1/ h2/h3 | Ara h9 | Ara h8 | Symptoms |
|---|---|---|---|---|
| + | + | + | - | Risk of Severe reactions |
| + | - | + | + | Mild reactions, cross-reactivity |

## CONCLUSION

Legumes such as soybeans and peanuts are among the most common food antigens that cause allergic reactions in North America and Asia. Previous reports have shown that in Spain, lentils are the legumes most commonly implicated in food allergy. Allergy to legumes may be mediated by primary sensitization via ingestion, or inhalation, or may be acquired after primary sensitization to pollen or to other foods. Allergic reactions to food may affect many different organs and systems. The gastrointestinal tract, including the oral mucosa and skin, are commonly involved. Airways are less commonly affected and the cardiovascular system may be affected in systemic anaphylactic reactions.

Between legumes, peanut is the food-allergen most implicated in the anaphylaxis series, and fatal reactions in food allergic patients. Besides, sensitization to others foods is frequent in peanut allergy patients, as sensitization and allergic symptoms with lupine. This fact could be relevant in the clinical course of allergic food patients.

In recent years, many legume´s allergenic proteins have been characterized and used within the context of molecular diagnosis or component-resolved diagnosis to facilitate more accurate diagnosis of food allergy. Most of the allergens already characterized from legumes are related with the 2S storage proteins superfamily. Regarding peanut, it has been demonstrated that peanut-allergic patients tend to recognize the major peanut components, Ara h 1-3. However, Ara h 2 offered the best discrimination between peanut-allergic and peanut-tolerant patients. In any case, the recognizition pattern might be variable depending on the region and the cooking methods.

## REFERENCES

[1] M. D. Ibáñez, M. Martínez, J.J, Sánchez and E.Fernández Caldas, *Allergol. et Immunopathol.* 31, 151 (2003).

[2] M. Martínez, M. D. Ibáñez and E. Fernández Caldas, Invest. *Allergol. Clin. Immunol.* 10, 187 (2000).

[3] N. Nicolau N and A. Custovic, *Curr. Opin. Allergy Clin. Immunol.* 11, 222 (2011).

[4] H. S. Skolnick, M. K. Conover-Walker, C. B. Koerner, H. A. Sampson, W. Burks and R. A. Wood, *J. Allergy Clin. Immunol* . 107, 367 (2001).

[5] S. L. Taylor, R. F. Lemanske, R. K. Bush and W.W. Busse, In: Chandra R. K., editor. *Food allergy.* St Johns: Nutrition Research Education Foundation. 21 (1987).

[6] C. Y. Pascual, J. Fernández Crespo, S. Sánchez Pastor, M. A. Padial, J. M. Díaz-Pena, F. Martín-Muñoz et al., *J. Allergy Clin. Immunol.* 103, 154 (1999).

[7] S. P. Patil, P. V. Niphadkar and M. M Bapat, *Ann. Allergy Asthma Immunol.* 87,140 (2011).

[8] C. Y. Pascual, J. F. Crespo, S. Sánchez-Pastor, R. Ayuso, G. García Sánchez and M. Martin Esteban, J. Ped. Neumol. 23, 41 (2001).

[9] M. L. Sanz, M. D. de las Marinas, J. Fernandez and P. M. Gamboa, Clinical Exp. Allergy. 40, 1461(2010).

[10] S. A. Bock, A. Muñoz-Furlong and H. A Sampson, *J. Allergy Clin. Immunol.* 119, 1016 (2007).

[11] K. A. M. B. Peeters, S. J. Koppelmann, A. H Penninks, A. Lebens, C. A. F. M. Bruijnzeel-Koomen, S. L Hefle et al., Allergy. 64, 549 (2009).

[12] D. A. Moneret-Vautrin, L. Gerin, G. Kanny, J. Flabbee, S. Fremont and M. Morisset, *J. Allergy Clin. Immunol.* 104, 883 (1999).

[13] A. Fiocchi, P. Sarratud, L. Terraciano, E. Vacca, R. Bernardini, E. Fuggetta, C. Ballabio et al. *Clin. Exp. Allergy.* 39, 1045 (2009).

[14] S. L. Hefle, R. F.Lemanaske and R. K. Bush, *J. Allergy Clin. Immunol* . 94, 167 (1994).

[15] D. De las Marinas, Z. Cojocariu, R. Escudero, N. Pardo and M. L. Sanz, J. *Investig. Allergol. Clin. Immunol.* 17, 227 (2007).

[16] B. Rojas-Hijazo, M. M. Garces, M. L. Caballero, P. Allonza and I. Moneo, Int. Arch. Allergy Immunol. 141, 47 (2006).

[17] A. Moreno-Ancillo, A. C. Gil-Adrados, C. Dominguez-Noche and P. M. Cosmes, Pediatr. *Allergy Immunol.* 16, 542 (2007).

[18] A. Prieto, E. Razzak, D. P. Lindo, A. Alvarez-Perea, M. Rueda and M. L. Baeza, *J. Investig. Allergol. Clin. Immunol.* 20, 76 (2010).

[19] J. Bernihisel-Broadbent and H. A.Sampson, *J Allergy Clin. Immunol.* 83, 435 (1989).

[20] C. Pascual, S. S. Pastor, S. De Miguel, G. G.Sánchez, J. Crespo and G. Salcedo, *J. Allergy Clin. Immunol.* 105, 133 (2000).

[21] M. Martínez San Ireneo, M. D. Ibáñez-Sandín, E Fernández-Caldas, F. Marañón Lizana, M. J. Rosales Fletes et al. *Int. Arch. Allergy Immunol.* 121, 137 (2000).

[22] R. Sanchez-Monge, G. Lopez-Torrejon, C.Y. Pascual, J. Varela, M. Martin-Esteban and G. Salcedo, *Clin. Exp. Allergy*. 11, 1747 (2004).

[23] G. Lopez-Torrejon, G. Salcedo, M. Martin-Esteban, A. Díaz-Perales, C.Y. Pascual et al. *J. Allergy Clin. Immunol.* 112, 1208 (2003).

[24] H. D. Belitz and W. Groch, In: Berlin, Springer-Velag 537 (1987).

[25] H. Breiteneder and C. Ebner, *J. Allergy Clin. Immunol.* 106, 27 (2001).

[26] D. D. Metcalfe, *Clin. Rev. Allergy*. 3, 331 (1985).

[27] R. K. Bush, S. L. Taylor and J. A. Nordlee, *Allergy Proc.* 10, 261 (1989).

[28] A. M. Scurlock and W. Burks, *Ann. Allergy Asthma Immunol.* 93, S12 (1993).

[29] B. Bohle, I. Swoboda and S. Spitzauer. Food antigens:structure and function. In: Metcalfe D. D., Sampson H. A., Simon R. A., editors. Malden, M. A. *Blackwell Science*, 38 (2003).

[30] International Union of Immunological Societies, *Allergen Nomenclature SubCommitte*, List of Allergens updated 2 December 2008.

[31] A.W. Burks, G. Cokrell, J. S. Stanley, G. A. Helm R. M., Bannon, J. *Clin. Invest.* 96, 1715 (1995).

[32] A. W. Burks, L. W. Williams, C. Connaughton, G. Cockrell, T. J. O' Brian, R. M. Helm, *J. Allergy Clin. Immunol.* 90, 962 (1992).

[33] W. Burks, H. A. Sampson and G. A. Bannon, Allergy. 53,725 (1998).

[34] A. E. Flinterman, E. von Hoffen, J. C. F. den Hartog Jager, S. Koppelman, S. G. Pasmans, M. O.Hoekstra et al. *Allergy.* 37, 1221(2007).

[35] *Allergen.* Avaliable at: http://www.allergen.org. Accesed October 1, 2010.

[36] G. A. Bannon, S. K. Li and H. A. Sampson, *J. Allergy Clin. Immunol.* 99, 141 (1997).

[37] A. J. Bircher, G. van Meller, E. Haller, B. Curty and P. C. Frei, *Clin. Exp. Allergy.* 24, 367 (1994).

[38] D. Mittag, J. Akkerdaas, B. K. Ballmer-Weber, L. Vogel, M. Wensing and W. M. Becker, *J. Allergy Clin. Immunol.* 114, 1410 (2004).

[39] A. Asarnoj, E. Ostblom, S. Ahsledt, G. Hedlin, G. Lilja and M. van Hage, et al. *Allergy*. 65, 213 (2010).

[40] I. Lauer, N. Dueringer, S Pokoj, S. Rehm, G. Zocatelli and G. Reese, et al. *Clin Exp. Allergy*. 39, 1427 (2009).

[41] S. Krause, G. Reese, S. Randow, D. Zennaro, D. Quarantino, P. Palazzo, et al. *J. Allergy Clin. Immunol.* 124,771 (2009).

[42] S. J. Maleki, O. Viquez, T. Jacks, H. Dodo, E. T. Champagne, S. Y. Chung et al. *J. Allergy. Clin. Immunol.* 112, 190 (2003).

[43] K. Beyer, E, Morrow, X. M. Li, L. Bardina, G. A. Bañón, A. W. Burks et al. *J. Allergy Clin. Immunol.* 107, 1077 (2001).

[44] W. C. Chiang, L. Pons, M. I. Kidon,W. K. Liew, A. Goh and W. Burks, *Pediatr. Allergy Immunol.* 21, 429 (2010).

[45] N. Nicolau, M. Poorafshar, C. Murray, A. Simpson, H. Winell, G. Kerry et al. *J. Allergy Clin. Immunol.* 125, 191 (2010).

[46] A. Vereda, M. van Hage and S. Ahlstedt, M. D. Ibañez, J. Cuesta-Herranz, J. van Odijk J et al. *J. Allergy Clin. Immunol.* 127, 603 (2011).

[47] A. Asarnoj, R. Moverare, E. Ostblom, M. Poorafshar, G. Lilja, G. Hedlin et al. *Allergy.* 65, 1189 (2010).

[48] B. Ballmer-Weber and S. Vieths, *Curr. Opin. Allergy Clin. Immunol.* 8, 270 (2008).

[49] H. B. Krishnan, W. S. Kim, S. Jang and M. S Kerley, *J. Agric. Food Chem.*;57, 938 (2009).

[50] M. Shibasaki, S. Suzuki, H. Nemoto and T. Kuroume. *J. Allergy Clin. Immunol.* 64, 259 (1979).

[51] H. P. Rihs, Z. Chen, F. Ruëff, A. Pettersen, P. Rozynek, H. Heiman et al., *J. Allergy Clin. Immunol.* 104, 1293 (1999).

[52] T. Holzhauser, O. Wackermann, B. K. Ballmer-Weber, C. Bindslev-Jensen, J. Scibilia, L. Perono-Garoffo et al. *J. Allergy Clin. Immunol.* 123, 452 (2009).

[53] D. Mittag, S. Vieths, L.Vogel, D. Wagner-Loew, A. Starker, P. Junziker et al. *J. Allergy Clin. Immunol.* 113, 148 (2004).

[54] K. Ito, S. Sjölander, S. Sato, R. Moverare, A. Tanaka, L. Soderstrom et al. *J. Allergy Clin. Immunol.* 128, 673 (2011).

[55] E. Guillamon, J. Rodriguez, C. Burbano, M. Muzquiz, M. M. Pedrosa, B. Cabanillas et al. *Mol. Nutr. Food Res.* 54, 1668 (2010).

[56] G. Zoccatelli, S. Pokoj, F. Foetisch, J. Bartra, A. Valero, M. del Mar San Miguel-Moncin et al. *Molecular Immunol.* 47, 1561 (2010).

[57] R. Sánchez-Monge, C. Y. Pascual, A. Díaz Perales, J. Fernández-Crespo, M. Martín-Esteban, and G. Salcedo, *J. Allergy Clin. Immunol.* 106, 955 (2000).

[58] D. Ibáñez Sandín, M. Martínez San Ireneo, F Marañón Lizana, E. Fernández Caldas, E. Alonso Lebrero and T. Laso Borrego, *Allergy.* 54, 1209 (1999).

[59] J. Akkerdaas, E. I. Finkina, S. A. Balandin, S. Santos Madagan, A. Knulst, M. Fernandez Rivas et al. *Int. Arch. Allergy Immunol*, 157, 51 (2012).
[60] R. Asero, G. Mistrello, D. Roncarolo, S. C. de-Vries, M. F. Gautier and C. L. Ciurana, et al. *Int. Arch. Allergy Immunol*. 122, 20 (2000).
[61] M. Wensing, A. C. Knulst and S. Piersma, F. O'Kane, E. F. Knol and F. J. Kopelman, *J. Allergy Clin. Immunol*. 111:4204 (2003).
[62] M. Radcliffe, H. Morrrow Brown, *The Lancet*. 365, 1360 (2005).
[63] M. M. B. W. Dooper, C. Plassen, L. Holden, H. Lindvik and C. K. Faeste, *J. Investig. Allergol Clin. Immunol*. 19, 283 (2009).
[64] R. A. Wood, *Paediatrics*. 111, 1631 (2003).

In: Peanuts: Bioactivities and Allergies ISBN: 978-1-62417-387-5
Editors: I.L. Expósito, A.B. Blázquez 

*Chapter 6*

# Peanut Allergy: Allergens and Etiology

***R. Jiménez-Saiz, E. Molina and I. López-Expósito****
Institute of Food Science Research (CIAL) CSIC-UAM, Madrid, Spain

## Abstract

Allergens from peanuts are labeled as class I food allergens, and they are mainly seed storage proteins that form part of the plant's defense system. They belong to the conglutin, vicilin and glycinin families. On the whole, peanut allergens are water-soluble glycoproteins stable against heat, acid and enzymatic digestion. The high number of peanut allergens makes it complex to study the etiology of peanut allergy that remains obscure, and although the pathway involved in allergic responses has been characterized, the underlying causes of the hypersensitive response are still under research. This chapter will review the biochemical and immunological properties of peanut allergens. A special attention will be put on Ara h 2 and Ara h 6, for they are considerably more potent than the rest. Regarding the etiology, firstly, the relationship between peanut exposure and the development of peanut allergy will be discussed. Furthermore, immunological mechanisms responsible for peanut allergy will be deeply described. These mechanisms include classical as well as new observations, such as the influence of the complement on peanut allergy etiology.

---

* E-mail address: ivan.lopez@csic.es.

## Introduction

Food allergies are immediate hypersensitive reactions to ingested foods that cause the release of mediators from intestinal mucosal and submucosal mast cells. Clinical manifestations are characterized by enhanced peristalsis, increased fluid secretion from intestinal lining cells and associated vomiting and diarrhea. Allergic reactions to food are also related to urticaria and eczema. Unfortunately, food-induced allergic reactions account for one-third to one-half of anaphylaxis cases in emergency departments worldwide, as well as fatal cases in the United States [1] and United Kingdom [2]. Eight foods are responsible for more than 90% of food allergies: cow´s milk, eggs, soy, wheat, peanuts, tree nuts, fish and shellfish. All food allergies have the potential to induce anaphylaxis, but without any doubt the peanut is the major one responsible for food-induced fatal reactions. Peanut allergy prevalence is increasing [3], it tends to present early in life, it does not usually resolve, and in highly sensitized people, trace quantities can induce allergic symptoms [4].

Peanuts contain approximately 25% protein. Between those, 11 allergens have been identified, named Ara h 1 to Ara h 11. The allergens in peanut have been studied intensively for more than 20 years, seeking to find a biochemical basis for the severity of peanut allergy and to explore whether protein modifications result in reduced allergenicity. However, modifications of peanut proteins may alter its botanic function, and it does not address the broad diversity of allergenic proteins that exist in peanut plants, where elimination of one will probably not eliminate the allergenic potential of other proteins. Nevertheless, certain modifications that can reduce or eliminate the allergenicity of some peanut allergens can induce an increase in others [5]. This chapter will extensively describe the biochemical and immunological properties of peanut allergens individually. The allergic reactions provoked by peanuts are strictly IgE-mediated type I hypersensitivity reactions. In such reactions, peanut specific IgE antibodies bind to high affinity receptors on mast cells and basophils. Later on, when peanut allergens penetrate mucosal barriers, cell-bound IgE and peanuts allergens crosslink, which results in degranulation of preformed allergic mediators and subsequent cell activation. These cells may then produce a variety of chemokines, which recruit other inflammatory cells and contribute to the IgE-mediated late-phase allergic response [6]. Although this is the main mechanism responsible for the symptoms observed in peanut-induced allergic reactions, it has been recently found that other pathways, such as the complement activation [7] or the mast cell-IgE-FcεRI-IL-13 pathway [8], might be implied in these reactions. An analysis of both classical as well as newly found immunological mechanisms related with peanut-induced reactions will be carried out on this chapter.

## PEANUT ALLERGENS

Peanut allergens are mainly seed storage proteins that form part of the plant's defense system. They belong to the conglutin, vicilin and glycinin families. On the whole, peanut allergens are water-soluble glycoproteins stable against heat, acid and enzymatic digestion. To date, there are a total of 11 peanut allergens, named Ara h 1 to Ara h 11, which have been identified and accepted by the Allergen Nomenclature Subcommittee of the International Union of Immunological Societies (IUIS) (Table 1). Ara h 1 to Ara h 8 are characterized by molecular weights between 12 kDa and 70 kDa and isoelectric points (pI) between 4.6 and 7.2. Classically, Ara h1, Ara h2 and Ara h3 were the peanut allergens considered immunodominant [9], but new findings postulate that Ara h 2 and Ara h 6 are the most potent and stable peanut allergens [10, 11], as well as point out the importance of Ara h 9, a lipid transfer protein (LTP) in the Mediterranean area [12, 13].

**Table 1. Peanut Allergens Identified Until April 2010**

| Peanut Allergen | Biochemical name | MW (kDa) |
|---|---|---|
| Ara h 1 | Cupin (Vicilin-type, 7S globulin) | 64 |
| Ara h 2 | Conglutin (2S albumin) | 17 |
| Ara h 3 | Cupin (Legumin-type, 11S globulin, Glycinin) | 60 |
| Ara h 4 | Cupin (Legumin-type, 11S, Glycinin) | 37 |
| Ara h 5 | Profilin | 15 |
| Ara h 6 | Conglutin (2S albumin) | 15 |
| Ara h 7 | Conglutin (2S albumin) | 15 |
| Ara h 8 | Pathogenesis-related protein, PR-10 | 17 |
| Ara h 9 | Nonspecific lipid-transfer protein 1 | 9.8 |
| Ara h 10 | 16 kDa oleosin | 16 |
| Ara h 11 | 14 kDa oleosin | 14 |

### Ara h 1

Ara h 1 is a glycoprotein of 63.5 kDa and pI of 4.55 that belongs to the 7S or vicilin family seed storage globulin [14]. It represents 12-16% of the total peanut protein [15]. Its structure presents a high number of mannose and complex *N*-glycans residues [16], including a signal peptide and a single glycosylation addition site (NAS: asparagine, alanine, serine) at amino acid position 521 to 523 [17]. So far, one isoform of Ara h 1 (Ara h 1.0101) has been reported.

Ara h 1 is a major peanut allergen recognized by over 95% of patients with peanut hypersensitivity, in which 23 different IgE-epitopes have been identified [18]. Moreover, Ara h 1 appears to be truncated at the N-terminal side in the plant upon maturation of the protein, and the cleaved-off region contains three allergenic epitopes, of which two are major [19]. Interestingly, Ara h 1 has been shown to activate dendritic cells and act as a Th2 adjuvant *in vitro* [20]. Going further, mutational analysis of the immunodominant epitopes revealed that single amino acid variations, especially in the center of each epitope, significantly decreased peanut-specific IgE binding and that hydrophobic residues seemed to be the most critical for IgE binding [21]. The position where hydrophobic amino acids are placed is very important in establishing Ara h 1 as an allergen. It is known that Ara h 1 forms large oligomers and most of the IgE binding epitopes are clustered at the contact points where each monomer joins the other, protected from protease degradation (Figure 1). This strategic placement of IgE binding epitopes on Ara h 1 contributes to its resistance to *in vitro* digestion [22] and probably to its resistance to thermal denaturation [23].

## Ara h 2

It is a glycoprotein of 17.5 kDa and a pI of 5.2 [24] with homology to the conglutin family. Ara h 2 represents 5.9-9.3% of total peanut protein content [15, 25] and two isoforms have been described (Ara h 2.0101 and Ara h 2.0201). The Ara h 2 structure includes 8 cysteine residues that could establish up to 4 disulfide bonds contributing to the overall protein stability and allergenicity. Ara h 2 can be purified from peanut seeds [26], but a reliable recombinant form of Ara h 2 has also been expressed in *E. coli* [27].

Ara h 2 is recognized by serum IgE from >90% of patients with peanut hypersensitivity [24]. It has been described as containing 10 linear IgE-binding epitopes within its structure, three of which (aa 27-36, 57-66 and 65-75) are recognized by the majority of the patients and are supposed to be immunodominant. Regarding the amino acids that form the epitopes, 63% are either polar uncharged or apolar residues. Interestingly, an *in vitro* functional assay to assess the capacity of Ara h 2 (among other peanut allergens) to degranulate humanized rat basophilic leukaemia cells that had been sensitized showed that 5-100 fold more molecules of Ara h 1 and Ara h 3 are necessary to obtain the same release as that obtained with Ara h 2 and Ara h 6, which points out Ara h 2 and Ara h 6 as the most potent peanut allergens [10]. When mutational analysis was conducted on Ara h 2 epitopes, it was seen that the single amino acid changes resulted in the loss of IgE-binding properties [25]. Moreover, T-cell-epitope mapping of Ara h 2 has been done and 4 immunodominant regions distributed throughout the linear sequence have been identified: 19-28, 45-53, 96-114 and 131-139. All T-cell epitopes mapped to different locations than the immunodominant IgE-binding sites, with the exception of one T-cell epitope: aa 19-28. Except for the last epitope, the rest T-cell epitopes would be suitable for therapeutic purposes. Besides its storage function, Ara h 2 was found to act as a trypsin inhibitor, which, in addition to being more resistant to trypsin digestion itself, was found to protect Ara h 1 from degradation by trypsin. It is known that roasting increases trypsin inhibitory activity of Ara h 2 approximately 3.5-fold [28]. Also, the enhancement of Ara h 2 allergenicity, attributed probably to neo-epitopes formation by Maillard reaction, is also provoked by roasting [29]. However, other heat treatments, such as boiling, did not affect IgE-binding capacity of Ara h 2 [30]. This fact and the extreme immunological stability of the core structures of Ara h 2 and Ara h 6 provides an explanation for the persistence of the allergenic potency even after food processing [31].

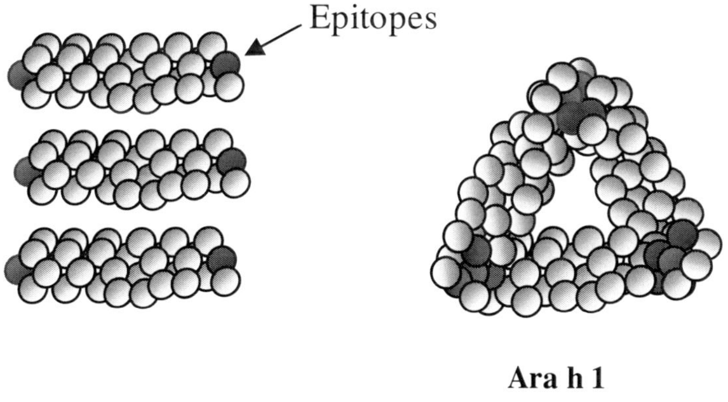


Figure 1. Placement of the epitopes of Ara h 1 within the subunits (left). These epitopes are encrypted when these chains join each other into the trimer (right).

## Ara h 3 and Ara h 4

Initially, Ara h 3 is synthesized as 60-kDa preglobulins [32] that are eventually cleavage, giving different polypeptides from approximately 14 to 45 kDa. After cleavage by an asparaginyl endopeptidase, the preglobulins become an $NH_2$-terminal acidic chain and a COOH-terminal basic chain, which later are linked by a disulfide bridge and then assembled, by head-to-head association, into hexameric oligomers from 300-450 kDa [33, 34]. The sequences of Ara h 3 and Ara h 4 are 91% identical, being isoallergens of the same gen,e but they were named independently [34, 35].

Both Ara h 3 and Ara h 4 have a pI of 5.5, and one isoform of each has been found (Ara h 3.0101 and Ara h 4.0101). Ara h 3 is considered a minor allergen that appears as a 14 kDa molecule on immunoblots with IgE serum from peanut-hypersensitive individuals [36]. By using a recombinant form of Ara h 3, which was recognized by about 45% of serum IgE from peanut allergic patients, 4 IgE-binding epitopes were identified in a study that screened 15 overlapping peptides derived from the Ara h 3 sequence. Similarly to Ara h 1 and Ara h 2, a single amino acid change on Ara h 3 epitopes could diminish IgE binding to 35-85% but keeps its capacity to induce T-cell activation, which could be of interest for immunotherapeutic aims [37].

Ara h 3 hexamers are reported to be stable to heat treatment with a denaturation temperature of 70-92ºC, depending on the conditions [38]. Looking at their structure, it seems that the epitopes are partially exposed on the surface of the native allergen and could be recognized after digestion because the IgE-binding sites would be more accessible [39]. Accordingly, Ara h 3 is not as stable as Ara h 2 or Ara h 6 to *in vitro* digestion [11].

## Ara h 5

Ara h 5 is a protein of 131 amino acids with a molecular weight of 14 kDa and pI of 4.6 [17]. Ara h 5 presents homology to plant profilins that are actin-binding proteins responsible for cytoskeleton formation in plant cells. Ara h 5 has been reported to have the isoform Ara h 5.0101, and its recombinant form has been expressed in *E. coli* and purified successfully [39, 40]. Ara h 5 is considered a minor allergen because of the low frequency of sensitization (13-16%), likely due to the small amount of Ara h 5 in peanut protein [42, 43]. Recently, 8 surface-exposed epitopes were predicted and verified, by checking IgE-binding with a microarray assay, in a recombinant model of Ara h 5, and 3 of them were found common among profilins [41]. This explains why Ara h 5 can cause cross-reactivity of Ara h 5 with profilins from birch or grass pollen, as well as wheat [40, 44, 45].

## Ara h 6 and Ara h 7

Ara h 6 is a protein of 124 amino acids that has a molecular weight of 14.5 kDa and a pI of 5.2. Ara h 7 has a molecular weight of 15.8 kDa and a pI of 5.6. Ara h 6 and Ara h 7 have a 35% amino acid sequence identical to each other. Both show similarities to amino acid sequences of proteins from the conglutin family, a seed storage family that conserved cysteine residues [42, 46].

The production of recombinant Ara h 6 and Ara h 7 and the purification of the natural forms has been described, although it was a difficult task due to the amount of peanut allergens with similar molecular masses that are accumulated in the range where natural Ara h 6 and Ara h 7 are expected as well as the availability of specific antibodies [47, 48].

In addition, one isoform of Ara h 6 (Ara h 6.0101) and two of Ara h 7 (Ara h 7.0101; Ara h 7.0201) have been reported.

Ara h 6 and Ara h 7 show 59% and 35% amino acid sequence identical to Ara h 2, respectively, which might explain why Ara h 6, based on its homology to Ara h 2, is considered a major allergen while Ara h 7 is not. However, more research is needed on Ara h 7, especially about the resistance to processing and digestion, to completely explore its allergenic potential, since recombinant Ara h 7 was recognized by immunoblot with a frequency of 43% of the patients sera in terms of IgE-binding.

The stability of Ara h 6 to *in vitro* digestion and processing has been studied, and it was reported that Ara h 6 could still be detected after 120 min of pepsin digestion and after 40 min of heating in the presence of sugar, butter, and wheat flour [47].

Ara h 6, like Ara h 2, contains cores that are highly resistant to proteolytic digestion and to temperatures up to 100ºC, and although IgE-binding capacity is reduced by protease treatment, the mediator released from the humanized rat basophilic leukaemia (functional equivalent of a mast cell or basophil) demonstrated that this reduction was not enough to decrease its allergenic potency [31].

## Ara h 8

Ara h 8 is a protein with 157 amino acid residues and a molecular weight of 16.9 kDa. It has an estimated pI of 5.03 [17] and two isoforms have been described (Ara h 8.0101 and Ara h 8.0201). The amount of Ara h 8 in peanut proteins is low, and, interestingly, it is not a seed storage protein like other peanut allergens (Ara h 1 - Ara h 7) but a member of the pathogenesis-related protein family PR-10. It might also be a steroid binding protein as it is Bet v 1, which shares 45.9% identity in their amino acid sequence with Ara h 8 and shows secondary structure similarities [49, 50]. Also, it has homology to Gly m 4 from soy. The natural Ara h 8 as well as its recombinant form has been extracted/expressed and purified successfully [50, 51]. It is considered a minor allergen but its great homology to Bet v 1, which is the major birch pollen antigen [52], makes it important for birch pollen allergic patients, because of the cross reactivity [50]. On the other hand, despite its homology to Gly m 4, the severe reactions that take place with Gly m 4 from soy have not been observed so far for Ara h 8. Regarding Ara h 8 stability, the effect of digestion and processing was studied, and it was reported that pepsin digestion completely destroyed the IgE-binding capacity of raw peanut extract within 5 seconds [50]. Moreover, IgE-binding capacity of the extract from roasted peanuts was 9-fold lower, suggesting low heat stability of Ara h 8.

## Ara h 9

Ara h 9 is a protein that belongs to the family of plant lipid transfer proteins with short chain (called $LTP_2$), which commonly have molecular masses from 9 to 10 kDa, present high pI, include 8 cysteine residues and are able to bind fatty acids and their derivates, as well as play a defensive role against pathogens [53]. The quantity of Ara h 9 in peanut proteins is supposed to be low, and, unlike other peanut allergens, its extraction requires acidic conditions. Recombinant forms of Ara h 9 have been expressed and two isoforms (Ara h 9.0101 and Ara h 9.0201) identified [12, 13].

The importance of Ara h 9 as a peanut allergen had been previously suggested [54], but it was confirmed recently [12]. For example, 19 (45.2%) sera of patients with peanut allergy (n=42) had IgE antibodies to Ara h 9, characterizing Ara h 9 as a minor allergen according to the rules of the IUIS. However, in the same study, only 4.8% of the subjects with peanut allergy showed IgE antibody reactivities to the classic allergens (Ara h 1, Ara h 2 and

Ara h 3), suggesting that no association between the IgE reactivity to the classic peanut allergens and the sensitization to Ara h 9 exists [12]. The frequency of peanut allergy in the Mediterranean area has been attributed, in a major extent, to Ara h 9 since peanut allergic individuals from the Mediterranean area (n=31) showed 90% prevalence of sensitization to the recombinant Ara h 9 isoforms (Ara h 9.0101 and Ara h 9.0201) [13]. Moving onto the stability of Ara h 9, it belongs to the LTPs, which are very stable with regard to digestion and heating. Such resistance to proteases and thermal processing is likely due to 4 α-helices forming a compact barrel-like structure stabilized by 4 disulfide bonds that highly contributes to its allergenic potential [54-56].

### Ara h 10 and Ara h 11

Ara h 10 and Ara 11 belong to the oleosin family and are low molecular mass amphipathic proteins with a long hydrophobic core, flanked by hydrophilic N- and C- terminal domains that take part of the plant´s lipid storage bodies (oil bodies) [57, 58]. Ara h 10 has a molecular weight of 16 kDa and 2 isoforms have been described (Ara h 10.0101 and Ara h 10.0102) while Ara h 11 has a molecular weight of 14 kDa and one isoform has been described (Ara h 11.0101) according to the IUIS. Both are recognized as minor allergens and since oleosins are oil soluble, they might account for reactions to peanut oil.

## Peanut Allergy Etiology

### Peanut: Exposure and Consumption

Sensitization to peanut—still a topic of discussion—may happen during pregnancy and lactation, depending on the mother´s diet. In this respect some clinical research has been done, but the results are contradictory regarding breastfeeding and development of peanut allergy in the offspring, although recent clinical studies suggest that including peanut within the maternal diet during pregnancy and lactation could increase the risk of sensitization to peanut in the newborn [59-61].

Once infancy is reached, the different degrees of environmental exposure to peanut are factors that can contribute to the development of peanut allergy or a tolerance state. Exposure to high levels of peanut by cutaneous contact and inhalation in childhood seems to promote sensitization, while low levels of exposure to peanut may be protective in atopic children [62]. Moreover, the presence of inflamed skin combined with the use of creams containing peanut oil might increase the risk factor for developing peanut allergy in children [62, 63].

The relationship of peanut consumption and the development of peanut allergy has been the focus of some studies aiming to understand the etiology of allergy to peanuts. An interesting research conducted in Jewish children from Israel and from the UK indicated that the prevalence of peanut allergy in Israeli Jewish children, who consume peanuts in high quantities in the first year of life, was lower than that of Jewish children in the UK, who avoid peanuts [59]. Moreover, animal investigations in rodents suggest that prior oral exposure to high doses of peanut can inhibit subsequent development of sensitization in terms of IgE antibody production [64, 65].

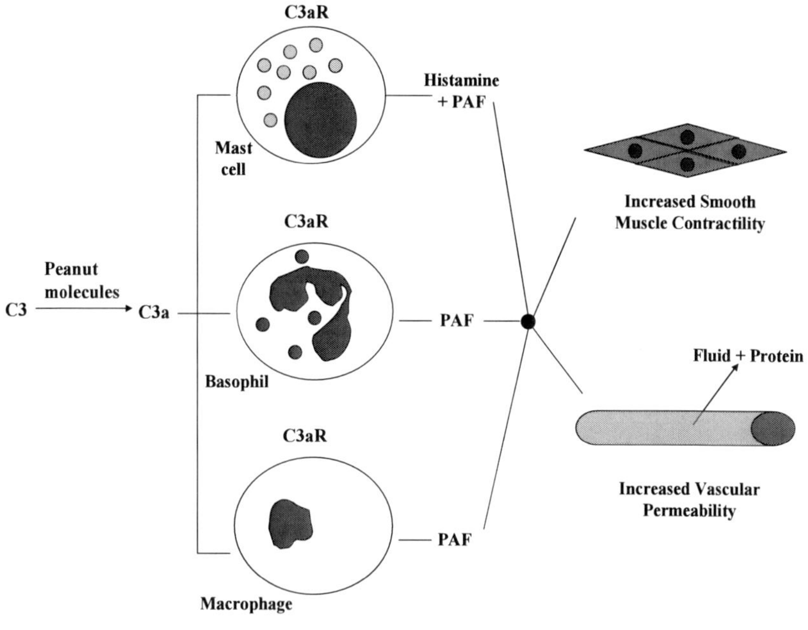


Figure 2. Peanut activation of complement exacerbates anaphylaxis. Peanut molecules activate complement, causing production of C3a, which acts through C3a receptors on mast cells, basophils, and macrophages to induce the production of PAF and histamine, which directly and indirectly increase vascular permeability and smooth muscle contractility. Adapted from [71].

Therefore these findings raised the question of whether early introduction of peanut in high amount during infancy, rather than avoidance, could be helpful in preventing the development of allergy to peanut, but the concept of "high" and "low" amounts, as well as the "early" and "late" introduction, should be clarified in human terms. In fact, some studies are underway, such as the Learning Early About Peanut Allergy (LEAP, www.leapstudy.co.uk) study, under the hypothesis that high-dose and early introduction of peanuts into children diets will induce oral tolerance and prevent the sensitization to peanut.

## Peanut Allergy: Immunological Mechanisms

Classically, the order of events in immediate hypersensitivity consists of exposure to an antigen, activation of Th2 cells and antigen-specific B cells, production of IgE antibody, binding of the antibody to Fc receptors on mast cells, and triggering of the mast cells by re-exposure to the antigen leading to the release of mediators from the mast cells and the subsequent pathologic reaction. However, new findings show that the immunological features of immediate hypersensitivity can differ depending on the food allergen. A recent study in a mouse model of peanut allergy suggested that the development of peanut-induced intestinal allergy is dependent on the mast cell-IgE-FcεRI-IL-13 pathway [66]. The requirements for IgE and mast cells are common in murine models of food allergy, but the role of IL-13 in peanut-allergy had not previously been shown. IL-13 has similar structure and function to IL-4 and works together with IL-4 in producing biologic effects related to allergic inflammation. Thus, IL-13 antagonist combined with IgE and mast cells antagonist could be helpful in counteracting peanut-allergic response. An interesting study conducted in mice demonstrated that, in contrast to milk and egg white, which have little ability to activate complement, peanuts extracts are able to activate complement and contribute to anaphylactic shock by causing production of C3a, which stimulates macrophages, basophils, and mast cells to produce platelet-activating factor and histamine (Figure 2) [67]. This ability to stimulate the complement might be the reason why peanut is involved in severe systemic reactions, due to the fact that complement is able to act as an adjuvant and therefore it contributes to the sensitization and elicitation phase of peanut allergy [68].

A very recent research, focused on the immune effector pathways contributing to peanut-induced anaphylactic reactions in mice, have shown that peanut-induced anaphylaxis is a process that involves the concerted action of multiple immune effector pathways, and thus interventions targeting a single pathway might not be enough to completely prevent anaphylactic responses. Mast cells, phagocytes and, to a lesser extent, basophils contributed to the peanut-induced systemic anaphylaxis in mice. The absence of any cell lineage prevented mortality and the concomitant deficiency of mast cells and phagocytes, but not mast cells and basophils, abrogated almost completely the anaphylactic responses. Moreover, mast cells were shown to contribute to anaphylaxis through IgE and IgG1, whereas phagocytes contribute through IgG1 exclusively [69]. Therefore, beside the classical pathway that leads to anaphylaxis mediated by IgE, FcεRI, mast cells, histamine and platelet-activating factor, there is another one mediated by IgG, FcγRIII, macrophages and platelet-activating factor [69, 70].

## Conclusion

Peanut allergy is considered a life-long allergy that offers a myriad of symptoms (including death) making it one of the most worrisome food allergies.

The major peanut allergens are Ara h 2 and Ara h 6 followed by Ara h 1, Ara h 3 and Ara h 7 and finally Ara h 8-11. These allergens are well characterized biochemically and structurally and their IgE-binding properties have also been described, but more detailed studies on the presence of certain epitopes in relation to the severity of peanut allergy are needed. Another aspect that remains to be elucidated is the influence of the food matrix on the immune responses. It is crucial to deepen the knowledge of how peanut allergens behave in the presence of other food components and how the allergenic potential is modified.

A deeper knowledge of peanut allergens, the immunological mechanisms involved in the allergic reaction and its clinical presentation are necessary to develop novel therapeutic and diagnostic approaches, due to the fact that dietary avoidance is the primary management of peanut allergy. Furthermore, the contribution of sensitization in utero, maternal diet or oral tolerance to peanut is still under discussion, and comprehensive studies are needed.

In the future, the use of purified allergens (from natural or recombinant source) can improve the ability in diagnosis and the cloning of peanut allergens may lead to the development of new strategies of immunotherapy and desensitization.

## ACKNOWLEDGMENTS

The authors want to acknowledge for financial aid to projects AGL2011-24740, CONSOLIDER-INGENIO2010CSDOOC-07-25506, CYTED/IBERO FUN 110AC0386 and COST ACTION FA1005.

## REFERENCES

[1] H. A. Sampson, *Pediatrics,* 111, 1601 (2003).

[2] J. O. Hourihane, S. A. Kilburn, P. Dean and J. O. Warner, *Clin. Exp. Allergy,* 27, 639 (1997).

[3] S. H. Sicherer, A. Munoz-Furlong, J. H. Godbold and H. A. Sampson, *J. Allergy. Clin. Immunol.* 125, 1322 (2010).

[4] S. Al-Muhsem, A. E. Clarke and R. S. Kagan, *Can. Med. Assoc. J.* 168, 1270 (2003).

[5] J. O. Hourihane, *Pediatr. Clin. North Am.* 58, 445 (2011).

[6] H. A. Sampson, *JAMA,* 278, 1888 (1997).

[7] M. Khodoun, R. Strait, T. Orekov, S. Hogan, H. Karasuyama, D. R. Herbert et al., *J. Allergy. Clin. Immunol.* 123, 342 (2009).

[8] M. Q. Wang, K. Takeda, Y. Shiraishi, M. Okamoto, A. Dakhama, A. Joetham et al., *J. Allergy. Clin. Immunol.* 126, 306 (2010).

[9] W. G. Shreffler, K. Beyer, T. H. T. Chu, A. W. Burks and H. A. Sampson, *J. Allergy Clin. Immunol.* 113, 776 (2004).

[10] F. Blanc, K. Adel-Patient, M. F. Drumare, E. Paty, J. M. Wal and H. Bernard, *Clin. Exp. Allergy,* 39, 1277 (2009).

[11] S. J. Koppelman, S. L. Hefle, S. L. Taylor and G. A. H. de Jong, *Mol. Nutr. Food. Res.* 54, 1711 (2010).

[12] S. Krause, G. Reese, S. Randow, D. Zennaro, D. Quaratino, P. Palazzo et al., *J. Allergy Clin. Immunol.* 124, 771 (2009).

[13] I. Lauer, N. Dueringer, S. Pokoj, S. Rehm, G. Zoccatelli, G. Reese et al., *Clin. Exp. Allergy,* 39, 1427 (2009).

[14] A. W. Burks, G. Cockrell, J. S. Stanley, R. M. Helm and G. A. Bannon, *J. Clin. Invest.* 96, 1715 (1995).

[15] S. J. Koppelman, R. A. A. Vlooswijk, L. M. J. Knippels, M. Hessing, E. F. Knol, F. C. van Reijsen et al., *Allergy,* 56, 132 (2001).

[16] D. Kolarich and F. Altmann, *Anal. Biochem.* 285, 64 (2000).

[17] H. W. Wen, W. Borejsza-Wysocki, T. R. DeCory and R. A. Durst, *Compr. Rev. Food Sci. Food Saf.* 6, 47 (2007).

[18] W. Burks, H. A. Sampson and G. Bannon, *Allergy,* 53, 725 (1998).

[19] H. J. Wichers, T. De Beijer, H. F. J. Savelkoul and A. van Amerongen, *J. Agric. Food Chem.* 52, 4903 (2004).

[20] W. G. Shreffler, R. R Castro, Z. Y. Kucuk, Z. Charlop-Powers, G. Grishina, S. Yoo et al., *J. Immunol.* 177, 3677 (2006).

[21] D. S. Shin, C. M. Compadre, S. J. Maleki, R. A. Kopper, H. Sampson, S. K. Huang et al., *J. Biol. Chem.* 273, 13753 (1998).

[22] S. J. Maleki, R. A. Kopper, D. S. Shin, C. W. Park, C. M. Compadre, H. Sampson et al., *J. Immunol.* 164, 5844 (2000).

[23] S. J. Koppelman, C. A. F. M. Bruijnzeel-Koomen, M. Hessing and H. H. J. de Jongh, *J. Biol. Chem.* 274, 4770 (1999).

[24] A. W. Burks, L. W. Williams, C. Connaughton, G. Cockrell, T. J. Obrien and R. M. Helm, *J. Allergy Clin. Immunol.* 90, 962 (1992).

[25] J. S. Stanley, N. King, A. W. Burks, S. K. Huang, H. Sampson, G. Cockrell et al., *Arch. Biochem. Biophys.* 342, 244 (1997).

[26] M. Sen, R. Kopper, L. Pons, E. C. Abraham, A. W. Burks and G. A. Bannon, *J. Immunol.* 169, 882 (2002).

[27] K. Lehmann, S. Hoffmann, P. Neudecker, M. Suhr, W. M. Becker and P. Rosch, *Protein Expr. Purif.* 31, 250 (2003).

[28] S. J. Maleki, O. Viquez, T. Jacks, H. Dodo, E. T. Champagne, S. Y. Chung et al., *J. Allergy Clin. Immunol.* 112, 190 (2003).

[29] S. J. Maleki, S. Y. Chung, E. T. Champagne and J. P. Raufman, *J. Allergy Clin. Immunol.* 106, 763 (2000).

[30] L. Mondoulet, E. Paty, M. F. Drumare, S. Ah-Leung, P. Scheinmann, R. M. Willemot et al., *J. Agric. Food. Chem.* 53, 4547 (2005).

[31] K. Lehmann, K. Schweimer, G. Reese, S. Randow, M. Suhr, W. M. Becker et al., *Biochem. J.* 395, 463 (2006).

[32] P. E. Staswick, M. A. Hermodson and N. C. Nielsen, *J. Biol. Chem.* 256, 8752 (1981).

[33] K. A. Barton, J. F. Thompson, J. T. Madison, R. Rosenthal, N. P. Jarvis and R. N. Beachy, *J. Biol. Chem.* 257, 6089 (1982).

[34] S. J. Koppelman, E. F. Knol, R. A. A. Vlooswijk, M. Wensing, A. C. Knulst, S. L. Heffle et al., *Allergy,* 58, 1144 (2003).

[35] H. W. Dodo, O. M. Viquez, S. J. Maleki and K. N. Konan, *J. Agric. Food Chem.* 52, 1404 (2004).

[36] P. A. Eigenmann, A. W. Burks, G. A. Bannon and H. A. Sampson, *J. Allergy Clin. Immunol.* 98, 969 (1996).

[37] P. Rabjohn, C. M. West, C. Connaughton, H. A. Sampson, R. M. Helm, A. W. Burks et al., *Int. Arch. Allergy Immunol.* 128, 15 (2002).

[38] E. L. van Boxtel, L. A. M. van den Broek, S. J. Koppelman and H. Gruppen, *Mol. Nutr. Food Res.* 52, 674 (2008).

[39] T. Jin, F. Guo, Y. W. Chen, A. Howard and Y. Z. Zhang, *Mol. Immunol.*. 46, 1796 (2009).

[40] T. Kleber-Janke, R. Crameri, S. Scheurer, S. Vieths and W. M. Becker, *J. Chromatogr. B,* 756, 295 (2001).

[41] C. Cabanos, M. R. Tandang-Silvas, V. Odijk, P. Brostedt, A. Tanaka, S. Utsimi et al., *Protein Expr. Purif.* 73, 36 (2010).

[42] T. Kleber-Janke, R. Crameri, U. Appenzeller, M. Schlaak and W. M. Becker, *Int. Arch. Allergy Immunol.* 119, 265 (1999).

[43] P. Maria, *Rom. Biotechnol. Lett.* 15, 5204 (2010).

[44] R. Valenta, M. Duchene, S. Vrtala, P. Valent, C. Sillaber, F. Ferreira et al., *Int. Arch. Allergy Immunol.* 99, 271 (1992).

[45] E. A. Pastorello, C. Incorvaia, V. Pravettoni and C. Ortolani, *Clin. Rev. Allergy Immunol.* 15, 415 (1997).

[46] S. J. Koppelman, G. A. H. De Jong, M. Laaper-Ertmann, K. A. B. M. Peeters, A. C. Knulst, S. L. Heffle et al., *Clin. Exp. Allergy,* 35, 490 (2005).

[47] M. Suhr, D. Wicklein, U. Lepp and W. M. Becker, *Mol. Nutr. Food Res.* 48, 390 (2004).

[48] H. Schmidt, S. Krause, C. Gelhaus, A. Petersen, O. Janssen and W. M. Becker, *J. Proteome Res.* 9, 3701 (2010).

[49] Z. Markovic-Housley, M. Degano, D. Lamba, E. von Roepenack-Lahaye, S. Clemens, M. Susani et al., *J. Mol. Biol.* 325, 123 (2003).

[50] D. Mittag, J. Akkerdaas, B. K. Ballmer-Weber, L. Vogel, M. Wenising, W. M. Becker et al., *J. Allergy Clin. Immunol.* 114, 1410 (2004).

[51] S. Riecken, B. Lindner, A. Petersen, U. Jappe and W. M. Becker, *Biol. Chem.* 389, 415 (2008).

[52] S. Flicker, S. Laffer, P. Steinberger, B. Alhani, Y. Zhu, M. L. Laukkanen et al., *Biol. Chem.* 381, 39 (2000).

[53] J. C. Kader, *Annu. Rev. Plant Physiol. Plant Mol. Biol.* 47, 627 (1996).

[54] R. Asero, G. Mistrello, D. Roncarolo, S. Amato, G. Caldironi, F. Barocci et al., *Allergy,* 57, 900 (2002).

[55] S. Scheurer, I. Lauer, K. Foetisch, M. S. M. Moncin, M. Retzek, C. Hartz et al., *J. Allergy Clin. Immunol.* 114, 900 (2004).

[56] A. I. Sancho, N. M. Rigby, L. Zuidmeer, R. Asero, G. Mistrello, S. Amato et al., *Allergy,* 60, 1262 (2005).

[57] A. H. C. Huang, *Annu. Rev. Plant Physiol. Plant Mol. Biol.* 43, 177 (1992).

[58] A. H. C. Huang, *Plant Physiol.* 110, 1055 (1996).

[59] G. Du Toit, Y. Katz, P. Sasieni, D. Mesher, S. J. Maleki, H. R. Fisher et al., *J. Allergy Clin. Immunol.* 122, 984 (2008).

[60] A. Des Roches, C. Infante-Riverd, L. Paradis, J. Paradis and E. Haddad, *J. Investig. Allergol. Clin. Immunol.* 20, 289 (2010).

[61] S. H. Sicherer, R. A. Wood, D. Stablein, R. Lindblad, A. W. Burks, A. H. Liu et al., *J. Allergy Clin. Immunol.* 126, 1191 (2010).

[62] A. T. Fox, P. Sasieni, G. Du Toit, H. Syed and G. Lack, *J. Allergy Clin. Immunol.* 123, 417 (2009).

[63] G. Lack, D. Fox, K. Northstone and J. Golding, *N. Engl. J. Med.* 348, 977 (2003).

[64] R. L. Thompson, L. M. Miles, J. Lunn, G. Devereux, R. J. Dearman, J. Strid et al., *Br. J. Nutr.* 103, 1278 (2010).

[65] I. Lopez-Exposito, K. M. Jarvinen, A. Castillo, A. E. Seppo, Y. Song and X. M. Li, *Food Res. Int.* 44, 1649 (2011).

[66] M. Q. Wang, K. Takeda, Y. Shiraishi, M. Okamoto, A. Dakharma, A. Joetham et al., *J. Allergy Clin. Immunol.* 126, 306 (2010).

[67] M. Khodoun, R. Strait, T. Orekov, S. Hogan, H. Karasuyama, D. R. Herbert et al., *J. Allergy Clin. Immunol.* 123, 342 (2009).

[68] C. W. Vandenberg, M. A. J. Hazenberg, F. M. A. Hofhuis, S. M. Vanrooyen and H. Vandijk, *Immunol.* 73, 264 (1991).

[69] K. Arias, D. K. Chu, K. Flader, F. Botelho, T. Walker, N. Arias et al., *J. Allergy Clin. Immunol.* 127, 1552 (2011).

[70] I. Lopez-Exposito, Y. Song, K. M. Jarvinen, K. Srivastava and X. M. Li, *J. Allergy Clin. Immunol.* 124, 1039 (2009).

[71] F. D. Finkelman, *Curr. Opin. Immunol.* 22, 783 (2010).

In: Peanuts: Bioactivities and Allergies ISBN: 978-1-62417-387-5
Editors: I.L. Expósito, A.B. Blázquez 

*Chapter 7*

# PEANUT ALLERGY IN THE MEDITERRANEAN AREA

***C. Mayorga[1], M. J. Torres[2] and A. B. Blázquez[1,*]***
[1]Research Laboratory, IMABIS Foundation-Carlos Haya Hospital
[2]Allergy Service, Carlos Haya Hospital, Málaga, Spain

## ABSTRACT

Allergy to peanut is becoming a significant health problem particularly in westernized countries. The prevalence of peanut allergy has increased and varies between 0.5-1.5%. The symptoms can be presented early in life and range from mild oral syndrome to anaphylactic reactions. Eleven peanut allergens have been described and recognized by the International Union of Immunological Societies (IUIS) Nomenclature Subcommittee named from Ara h 1 to Ara h 11.

No clear allergenic differences have been found between peanuts from different parts of the world. For example, in the United States, Spain, and Sweden, the peanuts consumed are typically dry roasted, on the other hand, in China, peanuts are eaten mainly boiled or fried. Perhaps this can be one of the reasons why patients around the world do not recognized the same allergens or react similarly. Data suggests that IgE response to Ara h 2 may be a useful marker for peanut allergy in the United States, Sweden, the U. K., and France, but it is unlikely an

* E-mail address: ablazquezgonzalez@yahoo.es.

important predictor of peanut allergy in Spain, where Ara h 9 appears to be a more relevant marker in this area.

## INTRODUCTION

Food allergy is an immunologically-mediated adverse reaction to food that encompasses a range of clinical disorders from acute to fatal anaphylactic reactions. Among these disorders, peanut allergy is becoming an important health problem, especially in westernized countries. To date, eleven peanut allergens have been described, with Ara h 1-3 being the most frequently involved in peanut allergy in the American population, whereas the non-specific lipid transfer protein (nsLTP) Ara h 9 seems to be more relevant in the Mediterranean area. One of the main properties of LTP proteins is their resistance to thermal or enzyme treatments, thus specific-IgE to LTPs is often associated with severe systemic symptoms.

The current management of food allergy is limited to strict dietary avoidance. Although attempts to desensitize patients with food allergy date back more than 100 years, there are no accepted therapies proven to accelerate the development of oral tolerance or to provide effective protection from unintentional exposures.

## MECHANISMS OF FOOD ALLERGY

The main site of exposure to food allergens is the gastrointestinal tract (GI), although other sites such as the skin or the respiratory tract may play a role in the sensitization to food proteins [1, 2]. To facilitate absorption and digestion of nutrients, the intestine is composed of a large surface area with a single layer of columnar epithelial cells forming the barrier between the lumen and the internal environments. The gastrointestinal-associated lymphoid tissue (GALT) is comprised of the three main categories of lymphoid tissue: immune cells distributed diffusely throughout the lamina propria (LP) or within the epithelium layer (IEL); the organized lymphoid nodules found within the Peyer´s patches (PP) and rectum, and the draining lymph nodes of the small and the large intestine (mesenteric lymph node (MLN) and sacral lymph nodes). Overlying the PP are specialized epithelial cells, termed *microfold* or *M cells*, which are efficient antigen sampling cells. Particulate antigens are preferentially taken up by M cells and can be presented by subepithelial dendritic cells to T cells within the PP [3, 4]. In contrast, soluble antigens can travel across enterocytes intact and gain access to immune cells in the LP, where they are taken up by dendritic cells (DCs) and transported to the draining mesenteric lymph node (MLN) [5, 6]. There, DCs present the antigen to naïve T cells which differentiated into several T cell subsets including Th1, Th2, Th17 or regulatory T cells producing a different pattern of cytokines and chemokines.

Antigens delivered via the oral route normally trigger a regulatory or tolerogenic immune response, also termed *oral tolerance* [7]. Although oral tolerance was initially defined in rodents by showing a systemic non-responsiveness after antigen feeding, this mechanism has also been found in human beings [8, 9]. A defect in the generation of this normal regulatory response to food antigens is known as food allergy. The immune mechanisms underlying food allergy or any allergic disease can be divided into two main steps: the sensitization and memory phase, and the effector phase. During the sensitization phase, the differentiation and clonal expansion of allergen-specific $CD4^+$ Th2 cells producing Th2 cytokines is essential for the induction of B-cell class-switching and the production of allergen-specific IgE antibodies. Allergen-specific IgE binds to the high-affinity IgE receptor (FcεRI) on the mast cells and basophils surface leading to the sensitization. During this step, there is generation of memory allergen-T and B cells. The effector phase initiates when a new encounter with the allergen leads to the cross-linking of the IgE-FcεRI complexes on sensitized mast cells and basophils, thus triggering their activation and release of mediators (histamine, prostaglandins, leukotriens…) responsible for the symptoms of the immediate phase of the type I hypersensitivity reactions. Continuous allergen exposure may trigger antigen-specific Th2 cell activation and production of Th2 cytokines, such as IL-13, IL-4, IL-5, and IL-9, which play a critical role in the maintenance of allergen-specific IgE levels, recruitment of inflammatory cells to the damaged tissue, mucus production, eosinophilia [10], and eventually systemic anaphylaxis.

## Prevalence of Peanut Allergy

Food allergy is a worldwide health problem, and peanut is one of the most allergenic vegetable foods. Peanuts are the most common food-related cause of IgE-mediated allergic reactions, particularly in western countries [11]. The reactions to peanut can present early in life and only a minority of peanut allergic people outgrow their allergy [12], thus peanut is one of the most common foods that cause an allergic reaction both in children and adults.

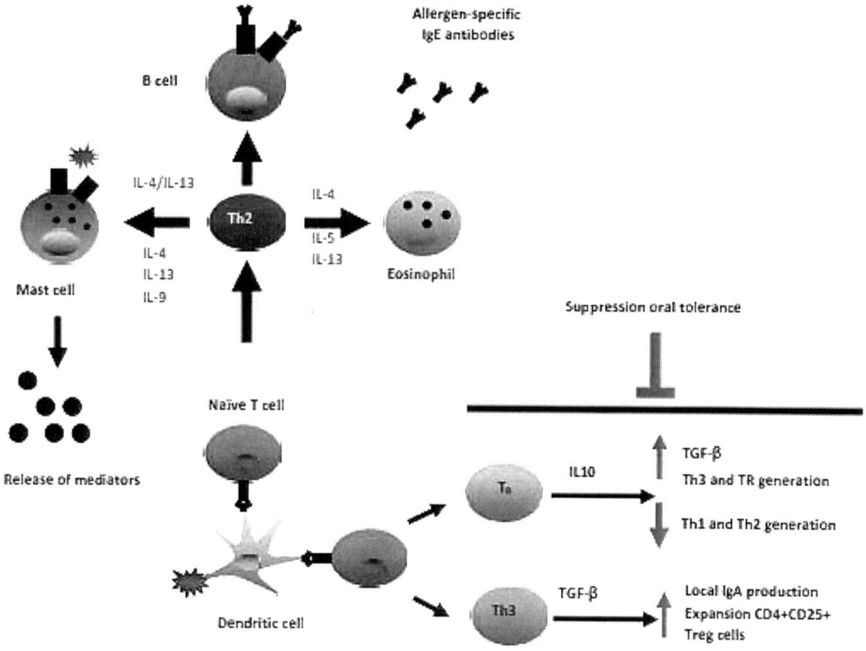


Figure 1. Mechanisms of food allergy. The initial sensitization to food allergens is thought to be through the GI track. Food allergens are absorbed and taken up by lamina propria DCs and presented to naïve T cells in the draining mesenteric lymph node. Primed T cells produced a wide range of Th2 cytokines leading to activation of B cells that eventually mature into IgE producing plasma cells. Antigen-specific IgE binds to the high affinity IgE receptor (FcεRI) expressed on the surface of mast cells and basophils. When the allergen is reintroduced, the IgE-FcεRI complexes cross-link leading to the activation and release of proinflammatory mediators, and eventually the food allergic response occurs.

Allergy to peanuts is common and likely increasing in prevalence. Peanut allergy is most commonly diagnosed between 6-24 months of age. Sicherer and colleagues reported that the prevalence of peanut allergy in children under 18 years of age increased from 0.4% in 1997 to 0.8% in 2002 in the United States [13]. Grundy et al [14] showed that sensitization to peanut increased between 1989 and 1994-96 from 0.5% to 1% in children 3-4 years of age in the U.K. And recently, Sicherer and collaborators, in an 11-year-follow-up study, reported that the prevalence of peanut allergy in children under 18 years of age had increased to 2.1% in 2010 [15].

Worldwide, the prevalence of peanut allergy is variable and the highest rates (1-2%) are seen in westernized countries, especially in the United States, Canada, and Australia. However, these rates may have stabilized in the U.K. and Canada [16, 17]. In contrast, the rates are lower in other westernized countries: France 0.3-0.7% [18], Israel 0.04-0.17% [19], and Denmark 0.2-0.6% [20]. These differences in the prevalence of peanut allergy are not only epidemiologic or clinical, but also have a molecular basis, perhaps because of frequency of ingestion of peanuts as well as method of preparation. For example, in Asia, the most common method of preparation of peanuts is boiled versus roasted in the United States [21]. The high temperature roasting (180 °C) leads to a Maillard reaction that appears to increase protein stability and thus the allergenicity [22, 23]. Emulsification (peanut butter) increases allergenicity through an adjuvant effect [24]. It has been shown that the manner in which foods are ingested might be relevant. For example, recent studies demonstrated that 70-80% of children allergic to milk or egg, tolerated baked (heat-denatured) but not unbaked forms of the protein [25, 26]. These results suggest that in these children, IgE antibodies recognized conformational epitopes and represent a group of children who naturally outgrow their food allergies.

Several studies have shown the prevalence of peanut allergy in the United States and Canada and how this is changing. In the United States, 1.4% of the subjects reported peanut or tree nut allergy [15]. Although this percentage has not changed for adults, in children less than 18 years old increased from 0.6% in 1997 to 1.2% in 2002 and 2.1% in 2007. Regarding the hospitalization rates for peanut-related anaphylaxis in New York State, it showed more than a 4-fold increase from 1990 to 2006 for individuals younger than 20 years old [27]. A recent study indicated that the majority of children considered peanut-sensitized on the basis of standard tests (IgE and skin prick testing) do not have a positive response to oral peanut challenge [28]. They also showed that the peanut component Ara h 2 was the most important predictor of clinical allergy.

On the other hand, it has been shown that in sensitized individuals, re-exposure to a particular allergen causes allergic reactions. However, the mechanisms underlying the initial sensitization remain unknown. There are multiple factors that may be important in the development of tolerance versus sensitization to peanuts, including route of exposure or dose. For example, comparing Jewish infants in Israel or in the U.K., it has been reported that the prevalence of peanut allergy in Israeli Jewish children was 0.17% compared to 1.85% in Jewish children in the U.K. These differences may be explained by the fact that Israeli infants (8-24 months) routinely consume peanut (7.1 g) whereas in the U.K., Jewish infants do not consume them at all (0 g) [19]. A control study in the U.K. found that early oral exposure to peanut in infants with high environmental peanut exposure may contribute to a protective effect against the development of peanut allergy, and that there was no effect of maternal peanut consumption during pregnancy or lactation with the development of peanut allergy [29]. Both studies showed that peanut sensitization occurs as a result of environmental exposure. In contrast, in animal models, it has been reported that low-dose peanut exposure during pregnancy and lactation reduced offspring risk of first exposure peanut reactions, and reduced active peanut specific-IgE sensitization [30].

## SENSITIZATION PROFILES IN PEANUT ALLERGY

Beyer et al [31] using recombinant peptides from Ara h 1, 2, and 3 demonstrated differences in epitope recognition between peanut allergic patients compared with those who were sensitized, but tolerant. This indicated that determination of IgE-binding epitopes may provide an additional tool to diagnose peanut allergy [31]. Shreffler and collaborators [32], using the protein microarray-based immunoassay, confirmed that Ara h 1, 2, and 3 are major peanut allergens and discovered an additional important epitope of Ara h 1 and the recognition of a remarkable degree of patient heterogeneity. Astier et al [33] used skin prick tests with recombinant allergens (Ara h 1, 2, and 3) for the diagnosis of peanut allergy. They demonstrated that co-sensitization to more than one component is predictive of more severe reactions. Flinterman et al [34] have shown that Ara h 2 and 6 are the major peanut allergens recognized more frequently in children. The first Asian study suggested that peanut-sensitized Asian children respond to the same major peanut allergens (Ara h 1-3) as North American and European peanut-sensitized children [35]. In contrast with this study, Vereda et al [36] demonstrated clinical and immunologic differences among patients from three different geographical regions. They recruited patients in the United States, Sweden, and Spain. American patients frequently recognized the three peanut allergens, Ara h 1-3, while Spanish patients were less often sensitized to these three peanut allergens. They were more often sensitized to the lipid transfer protein (LTP), Ara h 9, which, in turn, was rare in the United States and Sweden. Swedish patients recognized Ara h 1-3 more commonly than Spanish patients, but not as often as those from the United States, and had the highest sensitization rate to the Bet v 1 homologue Ara h 8. Spanish and Swedish children became allergic to peanut at 2 years of age or later, whereas the American patients became allergic around 1 year old. American and Spanish children experience more severe reactions than those from Sweden. This study demonstrated the heterogeneity in the clinical and immunological phenotype of peanut allergy in different geographical regions, and that might be because of the dietary traditions, exposure to different environmental factors, suggesting the importance of the molecular diagnosis in order to understand the peanut allergic disorders.

Double-blind placebo-controlled food challenge (DBPCFC) is currently considered the gold standard for peanut allergy diagnosis. However, this procedure requires the hospitalization of patients, mostly children, in specialized centers because oral exposure to allergens may cause severe reactions. Thus, there is a study [37] where the authors measured the levels of IgE to peanut extract and recombinant peanut allergens Ara h 1, 2, 3, 6, 7, and 8 in three groups of patients enrolled at two different hospitals (patients with proven peanut allergy, pollen-sensitized subjects without peanut allergy, and control subjects without allergic diseases) in order to evaluate the diagnosis by using this simple in vitro test. All peanut allergic patients were sensitized to peanut extract, 96% to Ara h 2, 92% to Ara h 6, 75% to Ara h 1, 61% to Ara h 3, 43% to Ara h 7, and 40% to Ara h 8. Despite being tolerant to peanuts, the majority of pollen-sensitized subjects had IgE directed towards peanut extract and Ara h 8, confirming the low specificity of this test in the diagnosis of peanut allergy. Specific-IgE to Ara h 2 was the best predictor of clinical reactivity to peanuts with a cut-off level of 0.23 kU/l conferring 93% sensitivity and 96% specificity. This study showed that a simple blood test can be used to diagnose peanut allergy with a high level of precision, however, double-blind placebo-controlled food challenge (DBPCFC) will remain useful for few cases to confirm the diagnosis.

Recently, a U.K. study [28] demonstrated that the measure of IgE to Ara h 2 was an even more useful predictor of clinical peanut allergy than the currently used skin or blood tests based on whole extract. In this study, the authors demonstrated that the majority of peanut-sensitized children do not have peanut allergy. They proceed to investigate the utility of a novel component-resolved diagnosis test using microarray technology amongst peanut-sensitized children. They found marked differences in the sensitization profile between peanut allergic and peanut tolerant subjects, resulting in a higher recognition of Ara h 1-3 by peanut allergic patients, whereas the peanut tolerant patients had higher values to cross-reactive carbohydrate determinants (CCDs) and grass components such as Phl p 1, Phl p 4, and Phl p 5b. This study demonstrated that Ara h 2 is the best marker to discriminate between peanut allergic and tolerant subjects. In a follow-up study, the value of specific IgE to Ara h 2 by ImmunoCAP has been shown, as a predictor of clinical reactivity to peanuts among peanut-sensitized school-age children in the U.K. [38].

Flinterman et al [39] used T-cell component-resolved diagnosis to differentiate between peanut-allergic and peanut-sensitized tolerant patients. They stimulated peripheral blood mononuclear cells (PBMCs) and short time T cell lines with crude peanut extract (CPE) and the purified peanut components Ara h 1-3, and 6. They found that T cell responses to CPE in peanut allergic and sensitized children were characterized by Th1 and Th2 cytokine production. However, only peanut-allergic children showed enhanced Th2 responses to Ara h 1, 3, and 6.

The data reported over recent years suggest that Ara h 2-specific IgE is a good predictor marker for peanut allergy in the United States [36], Sweden [36], the U.K. [28, 38], and France [37]. On the other hand, detection of Ara h 9 seems to be a more relevant allergen of peanut allergy in Spanish peanut allergic subjects [36, 40, 41].

## Lipid Transfer Proteins (LTPs)

Non-specific lipid transfer proteins (LTPs) are important allergens in fruits, nuts, vegetables, pollen, and latex. Despite their wide distribution throughout the plant kingdom, their clinical relevance is mainly confined to the Mediterranean area. LTPs can sensitize through the gastrointestinal route, thus they can be considered true allergens, and the IgE reactivity to LTPs is often associated with severe systemic symptoms. LTPs from different allergen sources due to structural homology are generally IgE cross-reactive. To date, the International Union of Immunological Societies Allergen Nomenclature Subcommittee lists 39 LTPs originating from fruits (n=18), pollen of trees and weeds (n=9), vegetables (n=7), nuts and seeds (n=4), and latex (n=1). LTPs are proteins of 9-10 kd, very stable to thermal and protease treatment, which is presumably a result of 4 $\alpha$-helices stabilized by 4 disulfide bonds. It has been shown that LTPs retain their allergenic activity during processing, in fact, even cooked food, fruit juices, jams, and fermented drinks still display IgE-binding activity [42]. In addition, several studies have demonstrated their resistance to pepsin proteolysis [43], gastric and duodenal digestion [44]. LTPs are potent elicitors of allergic reactions, including anaphylaxis, due to their physicochemical properties.

In the Mediterranean area, LTPs are major allergens from Rosaceae fruits, different vegetables, and nuts. In Northern and Central Europe, where birch trees are abundant, allergies to apple and cherry are related to food homologues of Bet v 1, the major birch pollen allergen. In contrast, in the Mediterranean area, allergy to Rosaceae fruits is related to peach allergy and sensitization to LTPs [45]. LTPs used to be concentrated in the pericarp of fruits, whereas the pulp contains lower amounts [46], explaining why some patients can tolerate fruits after peeling. Recently, it has been described that the LTP Ara h 9 as a major allergen in peanut-allergic patients from the Mediterranean area [41].

It has been observed a high IgE cross-reactivity for allergenic LTPs within the Rosaceae family [47], including the potent sensitizer molecule, Pru p 3 (the major peach allergen). Recently, allergenic LTPs from different sources (egg, mulberry [48], peanut [41], mustard [49], lettuce [50], green bean [51], or cabbage [52]) have been identified, and all of them show in vitro cross-reactivity with Pru p 3. Previous studies have shown that Pru p 3 is the most potent inhibitor molecule, while inhibition to Pru p 3 by others LTPs can only be partially achieved. It seems that Pru p 3 contain more epitopes and/or epitopes with higher IgE-binding affinity compared with other LTPs [43].

## TREATMENT

The primary treatment of food allergy is to avoid the causal food or foods in the diet. Education about avoidance includes careful attention in label reading, care in obtaining food from restaurants or food establishments, and avoidance of cross-contact of foods with the allergen during meal preparation, which makes dietary avoidance more difficult. Several medications can relieve symptoms of food allergic disorders. Antihistamines might partially relieve symptoms or oral allergy syndrome and IgE-mediated skin symptoms. Anti-inflammatory therapies are successful for allergic eosinophilic esophagitis or gastroenteritis [53]. Nowadays, a number of therapeutic strategies are under investigation targeting foods that most frequently induce severe IgE-mediated anaphylactic reactions. These approaches include both food allergen-specific and non-specific. Allergen-specific approaches include oral, sublingual, subcutaneous, and epicutaneous immunotherapy. Non-specific approaches include anti-IgE monoclonal antibodies, which might increase the threshold dose for reactivity to food allergens, and a Chinese herbal formulation, which reduced systemic anaphylaxis in a murine model of peanut-induced anaphylaxis [54] is currently being investigated in clinical trials.

Allergen-specific immunotherapy is clinically effective and it has been shown to prevent the development of asthma [55] and to reduce sensitization to other allergens [56]. In general, it is considered that 3 years of therapeutic treatment is optimal although this time depends on the allergen and the therapeutic preparation. Specific immunotherapy involves the administration of increasing doses of allergen extracts, which shows several disadvantages, such as severe anaphylactic side reactions [57] or leads to sensitization to new allergens present in the allergen mixture [58]. These mixtures contain whole antigen molecules with intact B-cell epitopes that are able to cross-link specific IgE molecules on the surface of effector cells and that may eventually trigger an anaphylactic reaction during the treatment. A number of strategies aimed at reducing the allergenicity of the allergen mixture used for the treatment. These approaches include conjugation with synthetic bacterial DNA motifs, chemical modification, point mutations in native allergen sequence, and the use of allergen derived B or T cell peptides [59-61]. Synthetic peptides containing T-cell epitopes show a reduced ability to cross-link allergen-specific IgE molecules [62]. Synthetic peptides have been developed and evaluated in experimental animal models and human clinical studies and this immunotherapy appear to induce a population of functional regulatory/suppressive Tr1 cells. In experimental mouse models, the induction of tolerance by the treatment with synthetic peptides depends on IL-10 production [63-65]. Similar mechanisms have been described for the peptide immunotherapy in human subjects [66, 67].

Subcutaneous immunotherapy has been used for more than 100 years to treat allergies. One study showed that peanut allergic patients treated with an aqueous extract of peanut a 67% to 100% decrease in symptoms during double-blind placebo-controlled food challenges (DBPCFC) and had a reduction in skin prick test reactivity to peanut at the end of the treatment, whereas 1 placebo-treated subject had no change in these parameters [68]. As a consequence of a pharmacy error, 1 placebo-treated subject died of anaphylaxis after the administration of a dose of peanut extract, resulting in the termination of the study. This tragic event and the fact that several studies provided that injected food allergen could induce unpredictable severe adverse reactions, highlighted the serious risks of this form of therapy.

Sublingual immunotherapy (SLIT) with food allergens represents another approach to desensitization and possible tolerance. The first report was described in a 29-year-old woman with a history of kiwi-induced anaphylaxis who received SLIT with fresh kiwi pulp extract [69]. In a double-blind placebo-controlled study, 18 children (1-11 years of age) were administered peanut SLIT or placebo as 6 months of dose escalation and 6 months of maintenance followed by a peanut DBPCFC [70]. This study showed that peanut SLIT is able to induce clinical desensitization in children with peanut allergy, with immunologic changes suggesting a significant change in the allergic response. A long-term study is required to determine whether continued peanut SLIT is able to induce immune tolerance. Preliminary data on food SLIT suggest a beneficial treatment effect, although significant adverse reactions are common.

An alternative route of allergen delivery is thought an epicutaneous (EPIT) patch. The first study was carried out in 18 children (mean age, 3.8 years; range, 10 months-7.7 years) with cow's milk allergy [71]. Children received three 48 h applications (1 mg of skimmed milk powder or 1 mg of glucose placebo) through the skin patch per week for 3 months. EPIT-treated children tolerated 23.6 ml (compared to 1.8 ml of milk that they tolerated at the initial point of the study) after 3 months of treatment; no changes were observed in the placebo group. There were no systemic reactions during the treatment. This small pilot study suggests that further investigation on EPIT for food allergy is required.

Oral immunotherapy (OIT) to food is one of the most actively investigated therapeutic approaches for food allergy, although few studies have established patient reactivity before treatment, included a placebo control or both. Furthermore, although most of the studies have shown the induction of desensitization in food allergic patients after OIT treatment, the development of tolerance has not been demonstarted. In addition to that, adverse reactions during treatment are common, >25% of doses are associated with adverse symptoms, although most are mild [72-75].

Since food allergy is an increasing problem in westernized countries, an effective therapy is required for this allergic disorder. Immunotherapy approaches now under study, attempts to avoid adverse effects, by changing the route of administration, doses of the food protein, or by engineering the proteins for the treatment.

A distinction must be made between desensitization, in which the allergen is ingested without symptoms during the treatment but requires daily ingestion, and tolerance in which the food protein might be ingested without symptoms despite periods of abstinence.

More studies are required to assess safety, efficacy, and mechanism. Most of these studies might address a better knowledge of the food allergen at the molecular level, route of administration, and adjuvants which improved the induction of tolerance.

## CONCLUSION

Allergy to peanut is becoming an important health problem around the world, and accurate diagnosis is critical. Diagnosis currently relies on a careful history, epidemiologic aspects, the role and limitation of simple diagnostic tests, and, if needed, the use of open food challenges to confirm allergy or tolerance. Ara h 2 may be a predictor marker of peanut allergy, especially in the U.K. and France; on the other hand, in the Mediterranean area, this allergen may not be a good prognostic marker being more relevant the LTP Ara h 9. Currently, treatment of food allergy is limited to the avoidance of triggers, and there is a medical need for an effective therapy for this allergic disorder. The exponential increase in research activity on food allergy and on different therapeutic strategies gives hope that an effective treatment for this allergic disease is within reach.

## REFERENCES

[1] Lack, G., Fox, D., Northstone, K., Golding, J. *N Engl J Med.* 348, 977 (2003).

[2] Strid, J., Hourihane, J., Kimber, I., Callard, R., Strobel, S. *Clin Exp Allergy.* 35, 757 (2005).

[3] Frey, A., Giannasca, K.T., Weltzin, R., Giannasca, P.J., Reggio, H., Lencer, W.I., Neutra, M.R. *J Exp Med.* 184, 1045 (1996).

[4] Neutra, M.R., Mantis,N.J., Kraehenbuhl, J.P. *Nat Immunol.* 2, 1004 (2001).

[5] Huang, F.P., Platt, N., Wykes, M., Major, J.R., Powell, T.J., Jenkins,C.D., MacPherson, G.G. *J Exp Med.* 191, 435 (2000).

[6] Liu, L.M., MacPherson, G.G. *Immunology.* 84, 241 (1995).

[7] Chehade, M., Mayer, L. *J Allergy Clin Immunol.* 115, 3 (2005).

[8] Husby, S., Mestecky, J., Moldoveanu, Z., Holland, S., Elson, C.O. *J Immunol.* 152, 4663 (1994).
[9] Kraus, T.A., Toy, L., Chan, L., Childs, J., Mayer, L. *Gastroenterology.* 126, 1771 (2004).
[10] Akdis, C.A., Akdis, M. *J Allergy Clin Immunol.* 123, 735 (2009).
[11] Boulay, A., Houghton, J., Gancheva, V., Sterk, Y., Strada, A., Schlegel-Zawadzka, M., Sora, B., Sala, R., van Ree, R., Rowe, G. *Allergy.* 63, 797 (2008).
[12] Skolnick, H.S., Conover-Walker, M.K., Koerner, C.B., Sampson, H.A., Burks,W., Wood, R.A. *J Allergy Clin Immunol.* 107, 367 (2001).
[13] Sicherer, S.H., Munoz-Furlong, A., Sampson, H.A. *J Allergy Clin Immunol.* 112, 1203 (2003).
[14] Grundy, J., Matthews, S., Bateman, B., Dean, T., Arshad, S.H., *J Allergy Clin Immunol.* 110, 784 (2002).
[15] Sicherer, S.H., Munoz-Furlong, A., Godbold, J.H., Sampson, H.A. *J Allergy Clin Immunol.* 125, 1322 (2010).
[16] Ben-Shoshan, M., Kagan, R.S., Alizadehfar, R., Joseph, L., Turnbull, E., St Pierre, Y., Clarke, A.E. *J Allergy Clin Immunol.* 123, 783 (2009).
[17] Venter, C., Hasan Arshad, S., Grundy, J., Pereira, B., Bernie Clayton, C., Voigt, K., Higgins, B., Dean, T. *Allergy.* 65, 103 (2010).
[18] Morisset, M., Moneret-Vautrin,D.A., Kanny, G. *Eur Ann Allergy Clin Immunol.* 37, 54 (2005).
[19] Du Toit, G., Katz, Y., Sasieni, P., Mesher, D., Maleki, S.J., Fisher, H.R., Fox, A.T., Turcanu, V., Amir, T., Zadik-Mnuhin, G.,Cohen, A., Livne,I., Lack, G. *J Allergy Clin Immunol.* 122, 984 (2008).
[20] Osterballe, M., Mortz, C.G., Hansen, T.K., Andersen,K.E., Bindslev-Jensen, C. *Pediatr Allergy Immunol.* 20, 686 (2009).
[21] Chung, S.Y., Butts, C.L., Maleki, S.J., Champagne, E.T. *J Agric Food Chem.* 51, 4273 (2003).
[22] Maleki, S.J., Viquez, O., Jacks, T., Dodo, H., Champagne, E.T., Chung,S.Y., Landry, S.J. *J Allergy Clin Immunol.* 112, 190 (2003).
[23] Beyer, K., Morrow, E., Li, X.M., Bardina, L., Bannon, G.A., Burks, A.W., Sampson, H.A. *J Allergy Clin Immunol.* 107, 1077 (2001).
[24] Sicherer, S.H., Sampson, H.A. *J Allergy Clin Immunol.* 120, 491 (2007).
[25] Nowak-Wegrzyn, A., Bloom, K.A., Sicherer, S.H., Shreffler, W.G., Noone, S., Wanich, N., Sampson, H.A. *J Allergy Clin Immunol.* 122, 342 (2008).

[26] Lemon-Mule, H., Sampson, H.A., Sicherer, S.H., Shreffler, W.G., Noone,S., Nowak-Wegrzyn, A. *J Allergy Clin Immunol.* 122, 977 (2008).

[27] Ben-Shoshan, M., Harrington, D.W., Soller, L., Fragapane, J., Joseph, L., St Pierre, Y., Godefroy, S.B., Elliott, S.J., Clarke, A.E. *J Allergy Clin Immunol.* 125, 1327 (2010).

[28] Nicolaou, N., Poorafshar, M., Murray, C., Simpson, A., Winell, H., Kerry, G., Harlin, A., Woodcock, A., Ahlstedt, S., Custovic, A. *J Allergy Clin Immunol.* 125, 191 (2010).

[29] Fox, A.T., Sasieni, P., du Toit, G., Syed, H., Lack, G. *J Allergy Clin Immunol.* 123, 417 (2009).

[30] Lopez-Exposito, I., Song, Y., Jarvinen, K.M., Srivastava, K., Li, X.M. *J Allergy Clin Immunol.* 124, 1039 (2009).

[31] Beyer, K., Ellman-Grunther, L., Jarvinen, K.M., Wood, R.A., Hourihane,J., Sampson, H.A. *J Allergy Clin Immunol.* 112, 202 (2003).

[32] Shreffler, W.G., Beyer, K., Chu, T.H., Burks,A.W., Sampson, H.A. *J Allergy Clin Immunol.* 113, 776 (2004).

[33] Astier, C., Morisset, M., Roitel, O., Codreanu, F., Jacquenet, S., Franck, P., Ogier, V., Petit, N., Proust, B., Moneret-Vautrin, D.A., Burks, A.W., Bihain, B., Sampson, H.A., Kanny, G. *J Allergy Clin Immunol.* 118, 250 (2006).

[34] Flinterman, A.E., van Hoffen, E., den Hartog Jager, C.F., Koppelman, S., Pasmans, S.G., Hoekstra, M.O., Bruijnzeel-Koomen, C.A., Knulst, A.C., Knol, E.F. *Clin Exp Allergy.* 37, 1221 (2007).

[35] Chiang, W.C., Pons, L., Kidon, M.I., Liew, W.K., Goh, A., Wesley Burks, A. *Pediatr Allergy Immunol.* 21, e429 (2010).

[36] Vereda, A., van Hage, M.,Ahlstedt, S., Ibanez, M.D., Cuesta-Herranz, J., van Odijk, J., Wickman, M., Sampson, H.A. *J Allergy Clin Immunol.* 127, 603 (2011).

[37] Codreanu, F., Collignon, O., Roitel, O., Thouvenot, B., Sauvage, C., Vilain, A.C., Cousin, M.O., Decoster, A., Renaudin, J.M., Astier, C., Monnez, J.M., Vallois, P., Morisset, M., Moneret-Vautrin, D.A., Brulliard, M., Ogier, V., Castelain, M.C., Kanny, G., Bihain, B.E., Jacquenet, S. *Int Arch Allergy Immunol.* 154, 216 (2010).

[38] Nicolaou, N.,Murray, C., Belgrave, D., Poorafshar, M., Simpson, A., Custovic, A. *J Allergy Clin Immunol.* 127, 684 (2011).

[39] Flinterman, A.E., Pasmans, S.G., den Hartog Jager, C.F., Hoekstra, M.O., Bruijnzeel-Koomen, C.A., Knol, E.F., van Hoffen, E. *Clin Exp Allergy.* 40, 590 (2010).

[40] Krause, S.,Reese, G., Randow, S., Zennaro, D., Quaratino, D., Palazzo, P., Ciardiello, M.A., Petersen, A., Becker,W.M., Mari, A. *J Allergy Clin Immunol.* 124, 771 (2009).

[41] Lauer, I., Dueringer, N., Pokoj, S.,Rehm, S., Zoccatelli, G., Reese, G., Miguel-Moncin, M.S., Cistero-Bahima, A., Enrique, E., Lidholm, J., Vieths, S., Scheurer, S. *Clin Exp Allergy.* 39, 1427 (2009).

[42] Zuidmeer,L., van Ree, R. *Curr Opin Allergy Clin Immunol.* 7, 269 (2007).

[43] Zuidmeer, L., Salentijn, E., Rivas, M.F., Mancebo, E.G., Asero, R., Matos, C.I., Pelgrom, K.T., Gilissen,L.J., van Ree, R. *Clin Exp Allergy.* 36, 666 (2006).

[44] Vassilopoulou, E., Rigby, N., Moreno, F.J., Zuidmeer, L., Akkerdaas, J., Tassios, I., Papadopoulos, N.G., Saxoni-Papageorgiou, P., van Ree, R., Mills, C. *J Allergy Clin Immunol.* 118, 473 (2006).

[45] Fernandez-Rivas, M., Bolhaar, S., Gonzalez-Mancebo, E., Asero, R., van Leeuwen, A., Bohle, B., Ma, Y., Ebner, C., Rigby, N., Sancho, A.I., Miles, S., Zuidmeer, L., Knulst, A., Breiteneder, H., Mills, C., Hoffmann-Sommergruber,K., van Ree, R. *J Allergy Clin Immunol.* 118, 481 (2006).

[46] Borges, J.P., Jauneau, A., Brule, C., Culerrier, R., Barre, A., Didier,A., Rouge, P. *Plant Physiol Biochem.* 44, 535 (2006).

[47] Garcia-Selles, F.J., Diaz-Perales, A., Sanchez-Monge, R., Alcantara, M., Lombardero, M., Barber, D., Salcedo,G., Fernandez-Rivas, M. *Int Arch Allergy Immunol.* 128, 115 (2002).

[48] Ciardiello, M.A., Palazzo, P., Bernardi, M.L., Carratore, V., Giangrieco, I., Longo, V., Melis, M., Tamburrini, M., Zennaro, D., Mari, A., Colombo, P. *Allergy.* 65, 597 (2010).

[49] Sirvent, S., Palomares, O., Vereda, A., Villalba, M., Cuesta-Herranz, J., Rodriguez, R. *Clin Exp Allergy.* 39, 1929 (2009).

[50] Hartz, C., Del Mar San Miguel-Moncin, M., Cistero-Bahima, A., Fotisch, K., Metzner, K.J., Fortunato, D., Lidholm, J., Vieths, S., Scheurer, S. *Mol Immunol.* 44, 2820 (2007).

[51] Zoccatelli, G., okoj, S., Foetisch, K., Bartra, J., Valero, A., Del Mar San MiguelMoncin, M., Vieths, S., Scheurer, S. *Mol Immunol.* 47, 1561 (2010).

[52] Palacin, A., Cumplido, J., Figueroa, J., Ahrazem, O., Sanchez-Monge, R., Carrillo, T., Salcedo, G., Blanco, C. *J Allergy Clin Immunol.* 117, 1423 (2006).

[53] Rothenberg, M.E. *J Allergy Clin Immunol.* 113, 11 (2004).

[54] Srivastava, K.D., Kattan, J.D., Zou, Z.M., Li, J.H., Zhang, L., Wallenstein, S., Goldfarb, J., Sampson,H.A., Li, X.M. *J Allergy Clin Immunol.* 115, 171 (2005).

[55] Jacobsen, L., Niggemann, B., Dreborg, S., Ferdousi, H.A., Halken, S., Host, A., Koivikko, A., Norberg, L.A., Valovirta, E., Wahn,U., Moller, C. *Allergy.* 62, 943 (2007).

[56] Pajno, G.B., Barberio, G., De Luca, F., Morabito,L., Parmiani, S. *Clin Exp Allergy.* 31, 1392 (2001).

[57] Bousquet, J., Lockey, R., Malling, H.J. *J Allergy Clin Immunol.* 102, 558 (1998).

[58] Moverare, R., Elfman, L., Vesterinen, E., Metso,T., aHaahtela, T. *Allergy.* 57, 423 (2002).

[59] Focke, M., Mahler, V., Ball, T., Sperr, W.R., Majlesi, Y., Valent, P., Kraft, D., Valenta, R. *FASEB J.* 15, 2042 (2001).

[60] Focke, M., Mahler, V., Ball, T., Kraft, D., Valenta, R. *Int Arch Allergy Immunol.* 124, 398 (2001).

[61] Moldaver, D., Larche, M. *Allergy.* 66, 784 (2011).

[62] Larche, M. *Allergy Asthma Clin Immunol.* 3, 53 (2007).

[63] Hoyne, G.F., O'Hehir, R.E., Wraith, D.C., Thomas, W.R., Lamb, J.R. *J Exp Med.* 178, 1783 (1993).

[64] Bauer, L., Bohle, B., Jahn-Schmid, B., Wiedermann, U., Daser, A., Renz, H., Kraft, D., Ebner, C. *Clin Exp Immunol.* 107, 536 (1997).

[65] von Garnier, C., Astori, M., Kettner, A., Dufour, N., Heusser, C., Corradin, G., Spertini, F. *Eur J Immunol.* 30, 1638 (2000).

[66] Alexander, C., Ying, S., Larche, M. *Clin Exp Allergy.* 35, 52 (2005).

[67] Tarzi, M., Klunker, S., Texier, C., Verhoef, A., Stapel, S.O., Akdis, C.A., Maillere, B., Kay,A.B., Larche, M. *Clin Exp Allergy.* 36, 465 (2006).

[68] Oppenheimer, J.J., Nelson, H.S., Bock, S.A., Christensen, F., Leung, D.Y. *J Allergy Clin Immunol.* 90, 256 (1992).

[69] Mempel, M., Rakoski, J., Ring, J., Ollert, M. *J Allergy Clin Immunol.* 111, 1406 (2003).

[70] Kim, E.H., Bird, J.A., Kulis, M., Laubach, S., Pons, L.,Shreffler, W., Steele, P., Kamilaris, J., Vickery,B., Burks, A.W. *J Allergy Clin Immunol.* 127, 640 (2011).

[71] Dupont, C., Kalach, N., Soulaines, P., Legoue-Morillon, S., Piloquet,H., Benhamou, P.H. *J Allergy Clin Immunol.* 125, 1165 (2010).

[72] Staden, U., Rolinck-Werninghaus, C., Brewe, F., Wahn, U., Niggemann, B., Beyer, K. *Allergy.* 62, 1261 (2007).

[73] Skripak, J.M., Nash, S.D., Rowley, H., Brereton, N.H., Oh, S., Hamilton, R.G., Matsui, E.C., Burks, A.W., Wood, R.A. *J Allergy Clin Immunol.* 122, 1154 (2008).

[74] Jones, S.M., Pons, L., Roberts, J.L., Scurlock, A.M., Perry, T.T., Kulis, M., Shreffler, W.G., Steele, P., Henry, K.A., Adair, M., Francis, J.M., Durham, S., Vickery, B.P., Zhong, X., Burks, A.W. *J Allergy Clin Immunol.* 124, 292 (2009).

[75] Blumchen, K., Ulbricht, H., Staden, U., Dobberstein, K., Beschorner, J., de Oliveira, L.C., Shreffler, W.G., Sampson, H.A., Niggemann, B., Wahn,U., Beyer, K. *J Allergy Clin Immunol.* 126, 83 (2010).

In: Peanuts: Bioactivities and Allergies ISBN: 978-1-62417-387-5
Editors: I.L. Expósito, A.B. Blázquez 

*Chapter 8*

# CURRENT UNDERSTANDING OF CROSS-REACTIVITY OF PEANUT AND POLLEN ALLERGENS

***C. Yang*[1,*] *and T. Zhang*[2]**

[1]Institute of Subtropical Agriculture, Chinese Academy of Sciences, Changsha, China

[2]State Key Laboratory of Food Science and Technology, Jiangnan University, Wuxi, Jiangsu, China

## ABSTRACT

Peanuts are one of the most important elicitors of food allergy in adults and adolescents. The allergic reactions towards peanuts have been frequently found in patients with birch pollen allergy. This correlation is mainly caused by the cross-reactivity of peanut and pollen allergens that share similar amino acid sequence identities. Comprehensive knowledge about cross-reactivity of peanut and pollen allergens is critical for the development of effective strategies to diagnose and treat allergic disorders associated with peanut and pollen and then to improve human health. Therefore, cross-reactivity of peanut and pollen allergens has long been of research interest. In the last decade, several cross-reactive allergens have been identified and characterized from peanut and pollen by molecular and immunobiochemical methods generating a vast amount

---

* E-mail address: chengbo1978@gmail.com.

of information in this area of research. The focus of this chapter is to critically review the current understanding of cross-reactivity of peanut and pollen allergens.

## INTRODUCTION

Peanuts are one of the most important elicitors of food allergy in adults and adolescents. Peanut allergy can induce all kinds of allergic reactions including life-threatening anaphylactic reactions [1, 2]. Symptoms usually develop within minutes after ingestion of even a trace amount of peanut and may involve cutaneous, cardiovascular, gastrointestinal, genitourinary, and/or respiratory systems [3]. It affects about 0.4-0.6% of children and 0.3-0.7% of adults with increasing prevalence in developed countries [4]. Peanut allergy is primarily mediated by allergen contact through the gastrointestinal tract. Up to 75% of children's allergic reactions to peanut occur at the first known exposure [5, 6]. However, more and more studies suggest that peanut allergy might also be acquired after a primary sensitization to birch pollen allergens on the basis of IgE cross-reactivity that is mainly caused by similar amino acid sequence identity of allergens [7, 8]. In accordance with the continuously increasing prevalence of allergic disorders caused by the cross-reactivity between peanut and pollen allergens, cross-reactivity of peanut and pollen allergens has become an important research area. In the last decade, several cross-reactive allergens have been identified and characterized from peanut and pollen by using molecular and immunobiochemical methods [7, 9]. There have been large amounts of research related to cross-reactivity of peanut and pollen allergens. It is beyond the scope of this chapter to summarize the vast information that has been generated in this area of research. The focus of the current chapter is to critically review current the understanding about the cross-reactivity of peanut and pollen allergens including major peanut and pollen allergens that have been isolated and characterized and major approaches to access cross-reactivity between peanut and pollen allergens.

## MAJOR PEANUT ALLERGENS

Major peanut allergens that have been isolated and characterized including Ara h 1, Ara h 2, Ara h 3, Ara h 5, Ara h 6, Ara h 7, Ara h 8, and Ara h 9 [8, 10-19]. Presently, most immunological, biochemical, and structural studies on peanut allergens have focused on the three major allergens (Ara h 1, 2, and 3) [10-12, 16]. Burks et al. (1991) indicated that Ara h 1 is a glycoprotein with a relative molecular weight of 63.5 kilodalton (kDa) and it has an isoelectric point of 4.55 [10]. Ara h 1 is a 7S vicilin-like globulin also known as conarachin [19]. Burks et al. (1991) demonstrated that Ara h 2 has two closely migrating bands with a mean molecular weight of 17 kDa [10]. Two-dimensional gel electrophoresis of this fraction revealed it to have a mean isoelectric point of 5.2. Ara h 2 is a 2S albumin, also known as conglutin, and functions as a trypsin inhibitor [20].

Rabjohn et al. (1999) [12] isolated a cDNA clone encoding a third peanut allergen, Ara h 3 and its amino acid sequence shows homology to 11S seed-storage proteins, and also functions as a trypsin inhibitor [21]. Peanut profilin (an actin-binding protein), Ara h 5, is one of the important peanut allergens and it comprises of 131 amino acid residues with a theoretical isoelectric point of 4.6, and a molecular weight of 14 kda [19]. Ara h 6 has been identified and exhibited characteristics of a food allergen regarding stability against digestion and thermal processing [17]. Natural Ara h 7 has been identified as the third peanut allergen of 2S albumin family [22]. Ara h 8 is a pathogenesis-related protein, also called PR-10 [23]. Ara h 9 has been identified as a new member of the lipid transfer protein (LTP) allergen family that seems to play an important role in peanut allergy for patients from the Mediterranean area [18].

## MAJOR POLLEN ALLERGENS

Birch pollen is a major cause of pollinosis and is responsible for cross-reactive oral allergies to fruits, nuts, and vegetables [24]. The major allergen, Bet v 1, has been extensively characterized, and six minor allergens, Bet v 2, Bet v 3, Bet v 4, Bet v 6, Bet v 7, and Bet v 8 have been cloned, sequenced [24-29], and included in the official allergen list of the WHO/IUIS allergen nomenclature subcommittee (http://www.allergen.org) [30]. Bet v 1, the major allergen of birch pollen, has been cloned and identified with a molecular weight of 17 kDa and it is a PR-10 protein [25, 27]. Seiberler et al. reported that Bet v 3 is not only a pollen allergen but also a $Ca^{2+}$-binding protein because of its unique structural features [28]. Vieths et al. (1997) indicated that Bet v 4 is a new class of pollen-specific $Ca^{2+}$-binding allergens with a physiological role as inhibitor of cytoplasmic streaming in outgrowing pollen tubes [35]. Cadot et al. (2000) has purified a novel allergen of birch pollen with a molecular weight of 18 kDa, Bet v 7, belonging to the cyclophilin family [29].

## Major Approaches to Measure Cross-reactivity

The cross-reactivity of two allergens can be measured qualitatively and quantitatively [7, 8, 31, 32]. The assessment of cross-reactivity between two allergens is currently primarily based linear sequence data as well as 3D information on homology among surface exposed residues [32]. Allergic reactions are mediated by allergen-specific IgE antibodies [31]. For quantitative assessment, three parameters are measured: 1) the fraction of the cross-reactive epitopes; 2) the fraction of the cross-reactive IgE; and 3) the relative affinity of the interaction between IgE and the two allergens [32]. RAST inhibition and double-blind, placebo-controlled food challenges (DBPCFC) have been widely used to access the cross-reactivity between two allergens [7]. T-cell activating capacity has been also used to evaluate cross-reactivity [37]. Gastrointestinal digestion does not destroy the T-cell activating capacity of pollen-related food allergens, and T-cell cross-reactivity occurs independently of IgE cross-reactivity [33]. Cross-reactive carbohydrate determinants is another distinct structure related to allergen cross-reactivity since many allergens are glycosylated and carbohydrate structure may also involve IgE binding [34].

## Cross-Reactivity of Peanut and Pollen Allergens

Bet v 1, Bet v 2, and Bet v 6 are known to have cross-reactivity with food allergens. Bet v 7 and Bet v 8 likely have IgE cross-reactions to homologous food proteins. Bet v 3 and Bet v 4 have no cross-reaction to food proteins [36]. The peanut profilin Ara h 5 has been shown to have cross-reactivity with the birch pollen profilin Bet v 2 since the cDNA-encoded gene product shows up to 80% amino acid sequence identity [9]. Ara h 8 has been indentified and cloned with 45.9% identity with the Bet v 1 amino acid sequence as well as shows IgE-reactivity, and Ara h 8 has low heat stability and can be easily digested leading to the immediate loss of their IgE-binding capacity [7]. These results suggest that successful specific immunotherapy (SIT) of pollen allergy should also inhibit allergic reactions to peanut. However, until now, there has been no report of the cross-reactivity between any other peanut allergens and pollen allergens. It has been shown that the cross-reactivity of peanut and pollen allergens is due to cross-reactive carbohydrate determinants [34].

## CONCLUSION AND PERSPECTIVE

Homologues of the birch pollen allergens Bet v 1 and Bet v 2 and glycoproteins carrying cross-reactive carbohydrate determinants have been indentified to play very important roles in the cross-reactivity of peanut and pollen. Further investigations, focusing on establishing the finest resolution of cross-reactivity of peanut and pollen allergens by using molecule-based approaches and bioinformatics approaches, are required. These new aspects regarding cross-reactivity between pollen and peanut allergens are very important for the development of efficient methods for the prevention and therapy of these allergic disorders.

## REFERENCES


[1] Bock, S.A., Munoz-Furlong, A., Sampson, H.A. *J. Allergy. Clin. Immunol.* 107, 191(2001).

[2] Wensing, M., Penninks, A.H., Hefle, S.L., Koppelman, S.J., Bruijnzeel-Koomen, C.A., Knulst, A.C. *J. Allergy. Clin. Immunol.* 110, 915(2002).

[3] Al-Ahmed, N., Alsowaid, S., Vadas, P. *Allergy. Asthma. Clin. Immunol.* 4, 139(2008).

[4] Sicherer, S.H., Munoz-Furlong, A., Sampson, H.A. *J. Allergy. Clin. Immunol.* 112, 1203(2003).

[5] Hourihane, J.O., Kilburn, S.A., Dean, P., Warner, J.O. *Clin. Exp. Allergy.* 27, 634(1997).

[6] Sicherer, S.H., Furlong, T.J., Munoz-Furlong, A., Burks, A.W., Sampson, H.A. *J. Allergy. Clin. Immunol.* 108, 128(2001).

[7] Mittag, D., Akkerdaas, J., Ballmer-Weber,B.K., Vogel, L., Wensing, Becker, W.M., et al. *J. Allergy. Clin. Immunol.* 114, 1410(2004).

[8] Mittag, D., Vieths, S., Vogel, L., Becker, W.M., Rihs, H.P., Helbling, A., et al. *J. Allergy. Clin. Immunol.* 113, 148(2004).

[9] Kleber-Janke, T., Crameri, R., Scheurer, S., Vieths, S., Becker, W.M. *J. Chromatogr. Biomed. Sci. Appl.* 756, 295(2001).

[10] Burks, A.W., Williams, L.W., Helm, R.M., Connaughton, C., Cockrell, G., O'Brien, T. *J. Allergy. Clin. Immunol.* 88, 172(1991).

[11] Burks, A.W., Williams, L.W., Connaughton, C., Cockrell, G., O'Brien, T.J., Helm, R.M. *J. Allergy. Clin. Immunol.* 90, 962(1992).

[12] Rabjohn, P., Helm, E.M., Stanley, J.S., West, C.M., Sampson, H.A., Burks, A.W., et al. *J. Clin. Invest.* 103, 535(999).

[13] Asero, R. *Ann. Allergy. Asthma. Immunol.* 83, 377(1999).

[14] Pons, L., Chery, C., Romano, A., Namour, F., Artesani, M.C., Gueant, J.L. *Allergy.* 57, 88(2002).

[15] Asero, R., Mistrello, G., Roncarolo, D., Amato, S., Caldironi, G., Barocci, F., et al. *Allergy.* 57, 900(2002).

[16] Koppelman, S.J., Knol, E.F., Vlooswijk, R.A., Wensing, M., Knulst, A.C., Hefle, S.L., et al. *Allergy.* 58, 1144(2003).

[17] Suhr, M., Wicklein, D., Lepp, U., Becker, W. *Mol. Nutr. Food. Res.* 48, 390(2004).

[18] Krause, S., Reese, G., Randow, S., Zennaro, D., Quaratino, D., Palazzo, P., et al. *J. Allergy. Clin. Immunol.* 124, 771(2009).

[19] Kleber-Janke, T., Crameri, R., Appenzeller, U., Schlaak, M., Becker, W.M.. *Int. Arch. Allergy. Immunol.* 119, 265(1999).

[20] Maleki, S.J., Viquez, O., Jacks, T., Dodo, H., Champagne, E.T., Chung, S.Y., et al. *J. Allergy. Clin. Immunol.* 112, 190(2003).

[21] Dodo, H.W., Viquez, O.M., Maleki, S.J., Konan. K.N. *J. Agric. Food. Chem.* 52, 1404(2004).

[22] Schmidt, H., Krause, S., Gelhaus, C., Petersen, A., Janssen, O., Becker, W.M. *J. Proteome. Res.* 9, 3701(2010); 9:3701-9.

[23] Riecken, S., Lindner, B., Petersen, A., Jappe, U., Becker, W.M. *Biol. Chem.* 389, 415(2008).

[24] Engel, E., Richter, K., Obermeyer, G., Briza, P., Kungl, A.J., Simon, B., et al. *J. Biol. Chem.* 272, 28630(1997).

[25] Breiteneder, H., Pettenburger, K., Bito, A., Valenta, R., Kraft, D., Rumpold, H.,et al. *EMBO. J.* 8, 1935(1989).

[26] Valenta, R., Duchene, M., Pettenburger, K., Sillaber, C., Valent, P., Bettelheim, P., et al. *Science.* 253, 557(1991).

[27] Swoboda, I., Jilek, A., Ferreira, F., Engel, E., Hoffmann-Sommergruber, K., Scheiner, O., et al. *J. Biol. Chem.* 270, 2607(1995).

[28] Seiberler, S., Scheiner, O., Kraft, D., Lonsdale, D., Valenta, R. *EMBO. J.* 13, 3481(1994).

[29] Cadot, P., Diaz, J.F., Proost, P., Van Damme, J., Engelborghs, Y., Stevens, E.A., et al. *J. Allergy. Clin. Immunol.* 105, 286(2000).
[30] Mahler, V., Fischer, S., Heiss, S., Duchêne, M., Kraft, D., Valenta, R. Int. Arch. Allergy. *Immunol.* 105, 5310(2001).
[31] Aalberse, R.C., Akkerdaas, J., Van Ree, R. *Allergy.* 56, 478(2001).
[32] Aalberse, R.C. *Clin. Mol. Allergy.* 5, 2(2007).
[33] Bohle, B. *Allergy.* 62, 3(2007).
[34] Guilloux, L., Morisset, M., Codreanu, F., Parisot, L., Moneret-Vautrin, D.A. *Int. Arch. Allergy. Immunol.* 149, 91(2009).
[35] Vieths, S., Lüttkopf, D., Reindl, J., Anliker, M.D., Wüthrich, B., Ballmer-Weber, B.K. *Allergy.* 57, 100(2002).
[36] Vieths, S. *Environ. Toxicol. Pharmacol.* 4, 61(1997).
[37] Fritsch, R., Bohle, B., Vollmann, U., Wiedermann, U., Jahn-Schmid, B., Krebitz, M., et al. *J. Allergy. Clin. Immunol.* 102, 679(1998).

# INDEX

## B



## C

## D

## E



## F

## G



## H



## I

## N



## O



## P

## Q



## R

## S

## T



## U



## V

## W